Adán Acosta-Morales
José Antonio Castán-Rocha
Salvador Ibarra-Martínez

Proposal for a NeuroFuzzy Model to Control Vehicle Traffic

Adán Acosta-Morales
José Antonio Castán-Rocha
Salvador Ibarra-Martínez

Proposal for a NeuroFuzzy Model to Control Vehicle Traffic

A focus on urban complexes

ScienciaScripts

Imprint

Cover image: www.ingimage.com

This book is a translation from the original published under ISBN 978-620-3-03108-9.

Publisher:
Sciencia Scripts
is a trademark of
International Book Market Service Ltd., member of OmniScriptum Publishing Group
17 Meldrum Street, Beau Bassin 71504, Mauritius
Printed at: see last page
ISBN: 978-620-3-39673-7

Summary

During the last decade, intelligent transport systems have been widely studied looking for contributions to different types of problems, being one of the most outstanding ones the efficient and effective management of urban roads. These contributions have been given through the implementation of different advanced computational techniques such as neural networks, genetic algorithms, multi-objective solutions, application of heuristic models, linear regressions, among others, achieving reliable and adequate results for the problems studied.

In spite of these contributions and their promising results at laboratory level, one of the questions that still remains very open is about the type of information that each author uses in his research; and not only the type but also the periodicity with which the information is obtained, the veracity of the data and above all the range with which each piece of information is classified. As we know, all advanced computing techniques base their procedures on the information they are offered to work with and the results will be as reliable and accurate as that information can be.

In this sense, one of the biggest problems faced by city administrators in recent years is that, within certain roadways, traffic regulation and control systems do not operate effectively because they are obsolete in the face of the challenges of this new technological era.

In light of this, we must pay special attention to the use of revolutionary techniques as a tool to solve static transport problems, generating an evolution towards cooperative systems, which should be flexible enough to be applied in dynamic environments. Such a statement is considered because the dynamics of a road environment is observed not as an aspect of movement but in relation to how road conditions change as time progresses, an effect that can easily be seen in the behaviour of an intersection. Therefore, today, some lines of implementation have emerged that promote the use of advanced information technologies for the design, study and analysis of road systems.

Keywords: Intelligent Transport Systems, Machine Learning.

Table of Contents

Introduction to Research.

1.1 Introduction

During the last decade, intelligent transport systems have been widely studied looking for contributions to different types of problems, being one of the most outstanding ones the efficient and effective management of urban roads. These contributions have been given through the implementation of different advanced computational techniques such as neural networks, genetic algorithms, multi-objective solutions, application of heuristic models, linear regressions, among others, achieving reliable and adequate results for the problems studied.

In spite of these contributions and their promising results at laboratory level, one of the questions that still remains very open is about the type of information that each author uses in his research; and not only the type but also the periodicity with which the information is obtained, the veracity of the data and above all the range with which each piece of information is classified. As we know, all advanced computing techniques base their procedures on the information they are offered to work with and the results will be as reliable and accurate as that information can be.

For example, we can find researches that use the number of vehicles on a road, which is obviously one of the most important data, however, some researches calculate this given as a series of time constants, others use finite state machines for the representation of the moments of car volumes, other authors perform vehicle gauging and develop random generators based on statistical models, therefore, on this basis, the results of the researches may not be comparable and/or may be understood as data that have not been previously evaluated.

The big problem is that the world's major cities demand that the services provided by governmental organisations work according to the demands of the population. An example of this is that the people who live in a given city require communications and transport services to work in a way that is appropriate to the demands posed by the challenges of population growth in each city from an individual perspective, since, in practice, one solution may not be appropriate for all cities, as each city develops in a specific and unrelated way. Therefore, city managers must be equipped with technologies that enable them to meet the above challenge adequately and efficiently.

In this sense, one of the biggest problems faced by city administrators in recent years is that, within certain roadways, traffic regulation and control systems do not operate effectively because they are obsolete in the face of the challenges of this new technological era.

On the other hand, it has been observed throughout the literature that when the situation of endless traffic literature is not well controlled, endless situations can arise that generate discomfort and dissatisfaction among users, among which we can comment:

- Motorists spend a great deal of time on their journeys (Gonzalez, 2003).
  - As a result, cars spend more time in operation, which constitutes higher fuel consumption and has a negative impact on the generation of greenhouse gases (Belikov et al., 2012), (Butler et al., 2012).
  - Pedestrians suffer from unsafe situations due to poorly planned pedestrian crossing times (Jaramillo, 2005), (Chaaban et al., 2005).
  - Public transport cannot respect a certain schedule due to traffic (Castan et al., 2012).

Derived from the above, it can be understood that vehicle traffic control in a city is, on many occasions, a problem that cannot be controlled and that cannot be prepared automatically to adequately resolve all the events that occur during the course of an average day. In this sense, the use of advanced computational techniques and methodologies have been used as a strategy capable of providing novel and effective solutions to solve some of the most common problems in traffic control, using the metaphor of autonomy as the main solution, well argued and demonstrated by the artificial intelligence community.

In fact, the SCT[1] in Mexico proposes a standard for measuring the quality of service at road intersections, which is an aid for traffic control experts to identify areas of opportunity for improvement, in the classic quest to offer users a better way to perform their transport operations. In that sense, intersections play an important role in the performance of a road network.

In addition, intersections are the bottlenecks of an urban network and critical factors of vehicle capacity, efficiency and safety. In addition, intersections are the bottlenecks of an urban network and critical factors in the vehicular capacity, efficiency and safety of an urban network. This is because an intersection comprises both the area where two or more streets meet (called a junction) and the entire space intended to facilitate the movements of vehicles travelling through it. In particular, according to (Garber et al., 88), the study of an intersection of two or more streets is a substantial ingredient in achieving the overall optimisation of vehicular flow in an urban network.

In light of this, we must pay special attention to the use of revolutionary techniques as a tool to solve static transport problems, generating an evolution towards cooperative systems, which should be flexible enough to be applied in dynamic environments. Such a statement is considered because the dynamics of a road environment is observed not as an aspect of movement but in relation to

[1] http://www.sct.gob.mx/

how road conditions change as time progresses, an effect that can easily be seen in the behaviour of an intersection. Therefore, today, some lines of implementation have emerged that promote the use of advanced information technologies for the design, study and analysis of road systems.

Thus, Intelligent Transport Systems (from now on ITS) appear as a new wave of tools that allow optimising, from different application perspectives, the vehicular flow. In fact, there are ITS that combine computational techniques with other disciplines such as: statistical control, prediction models, waiting lines, among others. In this direction, it has been possible to appreciate certain advantages when implementing a TIS within a road network, such as the following:

Improve road performance in terms of service volume.

- Decrease waiting times for vehicles at a red light.
- Therefore, the generation of gases is reduced.
- Avoid overwhelming congestion at peak hours.

Although the future of ITS is promising, the field of application remains a vision with a futuristic perspective. Several ITS applications and services are currently working effectively all over the world, generating a significant improvement in transport safety, mobility quality and productivity levels.

In this sense, the wide and deep application of ITS represents a real revolution in relation to the term mobility. New concepts, methods, tools and services continue to emerge in the areas of communication, information, automation and electronics, thus, ITS provide an important augmentation and an interesting platform for the field of Artificial Intelligence research and development, especially applying intelligent systems as solution tools. In this line of generation, several efforts have appeared that have allowed to evolve the way of designing, simulating and analysing road intersections.

However, due to some shortcomings of the existing tools, it is necessary to implement new instruments that allow traffic analysts to carry out studies on the levels of service that an arterial road (forming an intersection, whether simple or complex) could present and that at the same time are user-friendly, allowing ITS to consider real-time information and work in an autonomous way.

From the above arises the need for the necessary tools that allow traffic control experts to perform real-world tests on a tool that facilitates the integration of multiple factors and variables of a road environment, using methods, techniques and models that are well-founded and already well proven in the traffic control literature. Thus, the use of advanced computational technologies has emerged as a solution paradigm in the areas of intelligent transport systems, monitoring and optimising vehicle flow service levels.

1.2 Objectives.

The general objective introduces the methodological proposal put forward in this research to contribute to the state of the art in the field of intelligent transport systems. On the other hand, the specific objectives allow us to show the individual activities that will facilitate the fulfilment of the general objective of this research. In the following, we will present both objectives.

1.2.1 General Objective.

To develop a methodological proposal to optimise vehicular flow in urban areas through the implementation of a neuro-fuzzy model.

1.2.2 Specific Objectives.

- Identify which can be the main characteristics of a modern road and the possible ways to appreciate them.
- Establish a standardisation of the main variables used in the problem of urban vehicular traffic.
- To develop the neuro-fuzzy model for optimal urban traffic management.
- Carry out an experimental phase to test the statistical accuracy of the selected techniques.
- To present the main conclusions and contributions of the research carried out.

1.3 Justification.

During the last few years, research in the field of solving the problems and situations encountered in the search for intelligent vehicle traffic control has been tackled from different perspectives, with computer science being one of the most effective, presenting interesting results in the state of the art.

In this line we can find efforts dedicated to speed monitoring, traffic light management, priority systems for service cars, number plate detection, etc. It is important to mention that traffic is not the only problem that advanced computing is trying to solve. There are other well identified challenges such as: recommendation services, security, connectivity, information in the cloud, among others, which are an accurate reference that computer science, in any of its facets of usability, is a way to respond in an accurate, sustainable and effective way to solve problems in our times.

During the last years a lot of research has been dedicated to solve technological challenges that allow the functional growth of cities, through the use of cutting-edge technological tools, calling this avant-garde line "Smart Cities". It is important to mention that traffic is not the only problem to be solved, there are other well identified challenges such as: recommendation services, security, connectivity, information in the cloud, among others.

Proof of this is the current inertia of countries to develop "smart cities" that allow them to create integrated services to facilitate the way of life of the inhabitants and increase the quality of the processes that are carried out in these cities. The migration has not been and will not be easy, as it involves too many challenges, not only technological but also analytical and, above all, idiosyncratic. The latter seems almost impossible to exist, but despite the technological era in which we are currently immersed (most of the world's population has a telephone, a television, a refrigerator or simply a computer that uses intelligent technology or the internet) there are still services for which it has not been easy to create a technological proposal (e.g. automatic teller machines in banks that allow most of the services that were previously provided by a human advisor). Fortunately, what has been described above, rather than being seen as a problem, is being considered from the perspective of a design challenge or motivation. More and more researchers and technologists are designing and implementing their ideas to respond to the challenges described in the problem.

The information described above makes it evident that computer science has been and is an effective and efficient way to provide solutions to problems in different fields, where in recent years, vehicle traffic control has been one of the preferred fields of application for researchers due to the wide field of application and the great demand that this technology is experiencing in all countries of the world.

Finally, it should be considered that computer science is a current source of practical solutions in fields of real work application. For example, the so-called fourth industrial revolution reaffirms what has been described above, as we can find a great positive influence of computing and technology in its processes, which in their new conception require the use of the internet, advanced programming languages, robotics, cybersecurity, use and connection of devices, the cloud, among others.

1.4 Scope and Limitations.

1.4.1 Outreach

From the point of view of basic level research, establishing the paradigms that determine the scope of an investigation requires a perspective based on four axes of research methodology, which are exploratory, descriptive, correlational and explanatory, as presented in detail below:

1.4.1.1 Exploratory Scope.

This research is exploratory in scope since the problem being addressed is not entirely novel, but it presents diverse application approaches, which leads the authors to carry out a systematic search of the main contributions in the state of the art. According to the literature, there are indications of research efforts to improve the demand levels of a road using mathematical, computational and automated approaches. These previous works are the foundation and background that justify the efforts to carry out this research project.

1.4.1.2 Scope Descriptive.

The thesis argues preliminarily that it is necessary to adequately describe the way in which the problem is to be treated, since within a real road system there are different issues that dynamically modulate the behaviour of the road per se. This statement will be demonstrated once the background and the state of the art are presented, where it will be seen that each research uses different information and characterises it in different ways. Therefore, one of the first works to which the present research is oriented, is to establish what information will be used, to explain what this information is due to and how it could be conceptualised for its use within the computational models that are intended to be used.

1.4.1.3 Correlative Scope

Another defining issue of this research is the correlation of events on a road and how they can directly affect the way in which the management of such roads is determined. To better understand this, let us take as an example how a human operator regulates traffic when traditional traffic control systems (i.e., traffic lights) do not operate properly. A traffic officer decides manually and without a clear and precise system which cars should go first and this decision is made taking into consideration the current traffic situation on the streets. In this way we can understand that all the information that is presented on the road must be taken into account and must be studied with the necessary precision.

1.4.1.4 Explanatory Scope

Finally, this research tries to propose precisions that can determine the behaviour of cars on a road, but above all, it seeks to explain how traffic lights can behave to regulate road traffic under any situation at any stage of the device's useful life. In this way we will be able to understand how information seen from a practical perspective can lead to efficient and effective decision making.

1.4.2 Limitations

The limitations of this research are associated with two types of considerations, the methodological and the researcher's own, as explained below.

1.4.2.1 Methodologies

The research proposes a new paradigm to create intelligent systems in the field of transport with a focus on urban roads. Creating this proposal involves different methodological limitations, among which we can highlight the following:

- Define which aspects should be considered as important variables in the study of road performance. To achieve this, first of all, it is necessary to turn to experts in the field of traffic control, so that they can determine what information is really indispensable to try to improve the level of service of a road. Then it is necessary to turn to the literature base of the problem, selecting those experts in the field who can be useful for the scope of this research.

- Obtain current and real information on the behaviour of a group of roads in order to study and detect the phenomena that may occur. Normally such information is captured through in-situm surveys called vehicle counts, which count using different strategies (sometimes manually, sometimes using technology such as pressure probes). However, these data may not be entirely pure since the behaviour of a road from one hour to the next or from one day to the next may vary depending on the needs of the users. In this sense, it is proposed to conduct a study in at least five roads in the city of Tampico, Tamaulipas, Mexico in a period of 60 days during the months of October and November 2019. The sampling hours will be from 7 am to 7 pm.
- Show conclusive results demonstrating the effectiveness and efficiency of the proposal. The above is intended to be achieved by generating a system that allows the behaviour of a two-way road to be simulated, considering cars with different speeds and different types of driving. The aim is to generate a system as close to reality as possible in order to increase the level of reliability of the experiments and the results that can be obtained.
- To generate a level of confidence in relation to the data shown in this research, for which the authors confirm under oath that the data to be used will be random samples taken from real roads, which will not be altered for suitability. The data obtained will be made available to the state of the art.

1.4.2.2 From the researcher

The main challenge of this research is to demonstrate to city administrators that computer technology is a reliable and accurate tool that will allow adequate and robust traffic light control. However, obtaining the permits to carry out the sampling of the roads is another aspect of great consideration since at present the area of Tampico maintains a high level of security and therefore, to carry out the vehicular capacity must have the appropriate permits, which are intended to manage through the Faculty of Engineering "Arturo Narro Siller" FIANS of the Autonomous University of Tamaulipas UAT.

A further point to note is that the gauging will be carried out by the lead author of this research who will be supported by undergraduate students who are on secondment to the Robotics department at FIANS. This statement does not indicate in advance that the quality of the data will be compromised, however, there is a possibility that the sampling window at the times defined in the previous section may have to be adjusted.

In addition, in the absence of specialised equipment for the task described above, the challenge is to carry out the counts by means of manual counts supported by photographic images captured by the support group's personal mobile devices. Given this fact, there is the possibility that at some point there may be a counting deficit in the gauging. For this reason, the data will be studied using statistical tools.

On the other hand, in order to test the methodological proposal of this research, the author will develop a computational system that will replicate the behaviour of the variables of the roads, for which validation and adjustment tests will be carried out to compare the results of the proposed system against the data obtained in the vehicle registration.

On the other hand, we have the longitudinal effects, which have to do with the time needed to study the problem and measure possible changes in the behaviour of the performance measures that have been proposed within the scopes defined in the research.

Rationale and Background.

2.1 Fundamentals.

This section will be used to study the basic aspects of the topics of interest within the research, among which we can distinguish artificial intelligence, neural networks, fuzzy logic, intelligent transport systems and some aspects of smart cities.

2.1.1 Intelligent Transport Systems.

Intelligent Transport Systems (from now on ITS) appear as a new wave of tools that allow optimising, from different application perspectives, the vehicular flow. There are ITS that combine computational techniques with other disciplines such as statistical control, prediction models, waiting lines, among others. In this direction, it has been possible to appreciate certain advantages when implementing an ITS within a road network, such as:

- Improve road performance in terms of service volume.
- Decrease waiting times for vehicles at a red light.
- Therefore, the generation of gases is reduced.
- Avoid overwhelming congestion at peak hours.

Although the future of ITS is promising, the field of application remains a vision with a futuristic perspective. Several ITS applications and services are currently working effectively all over the world, generating a significant improvement in transport safety, mobility quality and productivity levels.

In this sense, the wide and deep application of ITS represents a real revolution in terms of mobility. New concepts, methods, tools and services continue to emerge in the areas of communication, information, automation and electronics, thus, ITS provide an important augmentation and an interesting platform for the field of Artificial Intelligence research and development, especially applying intelligent systems as solution tools. In this line of generation, several efforts have appeared that have allowed to evolve the way of designing, simulating and analysing road intersections.

According to the literature, there are several tools, models and methods to measure and simulate the behaviour on an arterial road. Particularly, this chapter is dedicated to present an overview of the most representative tools in the measurement, control and optimisation of vehicular flow, highlighting: TRANSYT-7F, SCOOT and SCATS.

The TRANSYT-7F system is a macroscopic model that considers a platoon of vehicles instead of individual vehicles. In particular, TRANSYT-7F simulates traffic flows at small incremental times. This representation of traffic is more detailed than other macroscopic models that assume a uniform distribution within traffic platoons (Wallace et al., 1991). The platoon dispersion algorithm simulates the spread of the platoon according to how the vehicles travel downhill, information which is also used by the model.

The TRANSYT-7F system simulates traffic for a single cycle and therefore no measure of effectiveness should be considered as an average for the analysis period. For the analysis, input information is required on: cycle lengths, data phases, approach volumes and turning movements, lane lengths, saturation flow rates and speeds (Zhang and Xie, 2006). Moreover, TRANSYT-7F calculates maximum queuing times. Such a calculation includes any vehicle joining the end of the queue after the signal changes to green and the front of the queue, considering the vehicle at the beginning of the queue.

The **SCOOT** (Split Cyclie Offset Optimization Technique) system is a tool designed to manage and control traffic signals in urban areas (Hunt et al., 1981). SCOOT is known as one of the leaders in urban traffic control, proven to reduce traffic delays by up to 20% in urban areas. SCOOT not only reduces delays and congestion problems, but also contains other traffic management advantages such as (Bretherton and Bowen, 1990):

Bus priority.

- Accident detection.
- Online saturation check measure.
- Estimation of vehicle emissions.

SCOOT is the world's leading adaptive traffic control system. SCOOT coordinates the operation of all traffic signals in an area to provide better progression on the road network (Bretherton, 1989). While coordinating all signals, SCOOT also responds intelligently and continuously to traffic flow changes and fluctuations throughout the day. Ultimately, the information provided by this tool allows optimisation of the use of vehicle control signals to minimise delays on the road network. As vehicles pass the detector, SCOOT receives the information and converts the data into internal units and uses this information to construct "flow cycle profiles" for each node.

The Sydney Co-ordinated Adaptive Coordinated Adaptive Traffic System (**SCATS**) is an intelligent transportation system developed in Sydney, Australia by the Roads and Traffic Authority in the 1970s and has been in use in Melbourne since 1982 (Bretherton and Bowen, 1990). SCATS has also been used in New Zealand, Hong Kong, Tehran, Dublin, Rzeszow and soon in part of Metro Atlanta, among other places. In summary, about 23,000 intersections in 100 countries use this system.

The Traffic Simulator is a tool widely used by traffic engineers to evaluate possible roadway modifications, alternative signalling configurations, new roadway construction and, in general, any type of innovation that has not yet been implemented. For the results to be correct, the predictions need to be accurate (Sims and Finlay, 1984). SCATS has been designed to detect changes in traffic flow and report changes in operating signals on a real-time basis, adjusting the signal settings (crossings, offsets and cycle times) at each cycle to compare trends in traffic flow and vehicle density (Sims and Dobinson, 1980).

Tables 1, 2 and 3 describe a detailed study of the main characteristics of a coordination model considering particular aspects of analysis.

Table 1. Optimisation Models in Signal Compensation.
Sequences, Cycle Length, Separations and Offsets

Model	Sequence	Cycle length	Separation	Compensation
TRANSYT-7F	Evaluate them entries based on the user specifications	Evaluaun rangopara each cycle with the best IP	Optimisation based on the best IP	Optimisation based on the best IP
SCOOT	Evaluates all that is allowed by a system of maximum bandwidth	Optimisation based on maximum bandwidth	Optimisation based on maximum bandwidth	Analysis Dynamic
SCATS	Evaluate them traffic and time signal settings	Optimisation based on the minimisation of downtime, delay and travel times.	Intersection-based optimisation	Default optimisation

Table 2. Optimisation Models in Signal Compensation. Applications, Criteria, Traffic Flow, Process Optimisation.

Model	Applications	Criteria/Flow of Traffic	Process Optimisation
TRANSYT-7F	Redesde Work	Uselessness (Delays and stops)/ Simulation Macroscopy	Using iterative gradient search to explicitly optimise the track spacing and compensation for the best given PI of the sequence phases and range of cycle lengths.
SCOOT	Redesde Work	Progression (breadthof band)/ Analytical	SCOOT obtains information about the traffic flow through the detectors; it not only reduces delays and congestion but also contains other traffic management facilities such as: Bus Priority, Traffic, Incident Detection, Online Saturation Measurement, Vehicle Emission Estimation, Information Capture Times.
SCATS	Intersections simple	Adaptive (breadthof band)/Inductive	SCATS allows for a hierarchical system of operational delay in each fault communication event. Such faults

			are monitored by the system; SCATS does not require extensive revisions to update timing sequences; SCATS constantly adjusts timing parameters to account for changes in traffic patterns, anticipating traffic flow details for planning proposals; SCATS provides reduction in the number of stops, reduces timing levels and saves significant fuel levels; SCATS provides reduction in the number of stops, reduces timing levels and saves significant fuel levels; SCATS provides reduction in the number of stops, reduces timing levels and saves significant fuel levels.

Table 3. Optimisation Models in Signal Compensation.
Strengths and Weaknesses of Models

Model	Fortress	Weakness
TRANSYT-7F	Able to optimise the signal amplitude of control networks; explicitly optimises the spacings and the offsets.	Unable to optimise sequencing steps; currently does not optimise cycle length; does not guarantee global optimisation; requires a solution
		initial; it has a macroscopic traffic flow model.
SCOOT	Identification of nodes that are causing congestion; Troubleshoot diagnostic problems where there are node failures; Report/diagnose problems where saturation rates are low; Diagnose and report where a node is not saturated; Diagnose and report where an intersection is not saturated; Diagnose and report where a intersection is overloaded; Pedestrian Facilities	Unable to optimise sequencing steps; currently does not optimise cycle lengths;
SCATS	Optimises pedestrian, public transport and general traffic flow; Responds effectively to changing traffic demand.	Unable to optimise sequence phases; Unable to optimise offsets and cycle lengths;

However, due to the fact that existing tools have some shortcomings, it is necessary to develop new instruments that allow traffic analysts to carry out studies on the levels of service that an arterial road (forming an intersection, whether simple or complex) could present and that at the same time are user-friendly, allowing ITS to consider real-time information and work in an autonomous way.

It is because of this fact, that the need arises for the necessary tools that allow traffic control experts to perform real-world tests on a tool that facilitates the integration of multiple factors and variables of a road environment, using methods, techniques and models that are well-founded and already well proven in the traffic control literature. Thus, the use of advanced computational technologies has emerged as a solution paradigm in the areas of intelligent transport systems, monitoring and optimising vehicle flow service levels.

Over the last decade, intelligent systems have gained a large foothold in the field of traffic control, demonstrating that they are a technological metaphor capable of efficiently and reliably solving the global challenges in the area of communications and transport services in any city.

Today, the saturation of transport infrastructures is mainly due to the increasing number of vehicles over the last forty to fifty years. The result of this increase in vehicle population is reflected in adverse situations such as traffic congestion, accidents of various kinds (i.e., collisions between vehicles, attacks on pedestrians, etc.), delays in public transport and one of the most relevant nowadays, the large amounts of pollutant emissions produced by cars. This situation affects our lives, particularly in urban areas, as people need to move more and more quickly and safely from one place to another.

As a first effort to solve these well-defined problems in the traffic environment, various solutions have been introduced that are capable of reducing these problems. Examples of these implementations are the use of common safety systems such as seatbelts or safety bags in vehicles, as well as the construction of more and better roads to facilitate the transport of citizens. However, the construction of new roads is not an adequate solution, as this represents a heavy investment and causes a negative impact on the environment, and it requires the existence of the necessary areas for such construction, which is an important limitation in residential areas.

On the other hand, it is easy to understand that it is essential to improve transport infrastructures in order to boost the economic development of a city. This is why the representatives and administrators of our cities are nowadays committed to satisfy such requirements by making use of the current trends in computer technology. Thus, the difficulties involved in such technological development have motivated the research community to focus its attention on the area of Intelligent Transport Systems (ITS). This area of study is oriented towards the scientific aspects whose main purpose is to develop new systems capable of solving some of the aforementioned problems.

It is important to clarify that the success of intelligent transport systems has depended on the results of different research activities exploited in different areas, such as electronics, control, communications, sensors, robotics, signal processing and information systems. This multidisciplinary nature increases the complexity problem of such systems, as, due to this, knowledge transfer and closer cooperation between different areas of knowledge enhancement is required.

Transport systems are at the heart of any nation's economic development. In this case, every developed or expanding city has generated transport systems that allow them to achieve efficient traffic control in both private urban traffic and bus and train services. Traffic monitoring and control is one of the key aspects, especially in metropolitan areas, as it takes into account both the number of vehicles that commonly increase year by year and pedestrians, who are, unfortunately, becoming more and more reckless.

Throughout the literature, there are several technological products capable of reducing travel delays and reducing emissions produced by cars. Some of this work involves the use of electronic controllers based on the use of microcontrollers and microprocessors. These controllers are commonly used in the traffic lights that we see today, but it is important to clarify that they present certain limitations because these traffic lights do not have the flexibility to modify their behaviour during the execution of their cycles.

In traffic control systems, the identification and consideration of the conditions of an intersection are two extremely important aspects as they are the basis of the data used to determine the timing of each signal. Accordingly, (Saeed et al., 2011) presents the implementation of a fully autonomous multi-agent system using a fuzzy logic model equipped with wireless sensors to solve problems such as congestion, accidents, speed and traffic irregularities. In fact, this approach provides a solution that minimises the waiting time of vehicles and even gives a higher priority to emergency cars by means of the fuzzy controller, trying to give a new approach to the services to be offered and facilitated by an intelligent infrastructure.

Also, in (Perez et al., 2010) a new infrastructure is presented that allows communication between vehicles and the control system to monitor the speed of cars. This approach uses radio frequency identification devices (RFID) so that traffic lights can recognise each of the cars on the road. In addition to the above, this approach implements the measurement of the speed of cars using a sensor based on the Hall effect. In order to achieve a more efficient adaptation of the car speed, the approach introduces a fuzzy logic approach using both the sensor information as well as the information obtained from the traffic light. Finally, the results suggest unexpected circumstances and improve the safety of car occupants.

Along the same route, a smart traffic light controller using an FPGA module with a neuro-fuzzy system is introduced in (Zade and Dandekar, 2011). This approach is able to make real-time decisions to reduce delays at intersections. The fuzzy logic theory in the controller provides the green light intervals in response to traffic conditions through the use of certain input variables that control the flow of heavy traffic at the study intersection and neighbouring arterials. This contribution argues that it is able to overcome the weaknesses of conventional traffic controllers by using its ability to provide each traffic light with a variety of green light intervals according to the car volume loads at a four-way intersection. In (Roozemond, 2001) an agent-based urban intersection control system was published, which reacts to changes in the road environment and adapts itself to the changes according to a set of predefined rules.

In addition to the current trends in solving various computational methods, the development in the field of vehicular traffic simulation environments is another interesting and promising field of study. Today, computational tools exist that offer the opportunity to implement and test approaches to corroborate the techniques, models and methodologies that are being developed at the experimental science level to cope with the demands of our cities. However, these tools are not functional for all proposals. Because of this, some authors tend to develop their own simulators while they are generating their methodological proposals.

For example, in (Huang et al., 2012) a reservation-based approach to control for road intersections was designed and evaluated. This paper proposes the development of a new simulator for testing in an interconnected vehicle environment. Indeed, the system integrates a microscopic traffic simulator with a road network simulator and an emission analyser. Some experiments are presented using the simulator which was developed with the main interest of comparing the vehicle mobility and environmental benefits of the proposed approach against traditional control methods. References (Chen et al., 2005; Chen et al., 2006; Garcia-Serrano et al., 2003; Hernandez et al., 2002; Liu et al., 2005; Wang, 2005; Wang, 2008; Zhanh et al., 2004) are other examples where the development of an auxiliary computational tool was necessary for the application of their methodological approaches.

To conclude this analysis of artificial intelligence-based intelligent transport systems, (Chen and Cheng, 2010) provides an interesting and comprehensive review of the literature on agent-based approaches, which effectively places the dominance of transport systems over any other effort to solve traffic problems. Indeed, one of the main advantages of ITS is that they are able to have a geographically distributed domain in dynamically changing environments. In particular, the authors review various agent-based traffic applications and encapsulate them in a five-category classification: 1) architectures and platforms for agent-based traffic control and management; 2) agent-based systems for road transport control; 3) agent-based systems for air traffic control and management; 4) agent-based systems for rail transport; and 5) traffic modelling and simulation using multi-agent systems.

Based on the above text, the use of agents and artificial intelligence in the development of urban traffic control systems emerges as an adequate and effective solution to avoid the collapse of communication and transport services in a modern city. Despite the above, it is important to mention that the design, development and implementation of agent technology in traffic systems is still an immature effort and needs to continue to grow in order to achieve a better level of accuracy, reliability and safety so that the construction of traffic control systems is more robust and can be implemented in a fully efficient way in the streets of any city in the world.

Some previous work has been devoted to assessing the mobility and environmental benefits of smart intersection systems. However, the fact that a smart intersection is dedicated to reducing the time that cars are unnecessarily stopped at red lights and acceleration and deceleration manoeuvres implies that intersections could also reduce fuel consumption, greenhouse gas emissions and the generation of other pollutants (e.g. brake dust, worn tyres, brake pads, etc.). From the above, we can surmise that some of the benefits of smart intersections have not been clearly quantified in the literature.

Some of the work on smart intersections is not able to utilise information about previous decisions that were made that were functional in solving a roadway problem under circumstances of traffic volume overload.

Other models that have been implemented for traffic light control are unable to use a communication or interaction protocol that allows traffic lights to reach a consensus to improve the synchronisation of their lights, in order for intersections to be truly autonomous and communicative, two of the main qualities of intelligent entities.

Simulators that are currently registered in the state-of-the-art cannot work in real time and, therefore, they cannot auto-synchronise the timing of the lights of each traffic light while an experiment is running.

2.1.2 Artificial Intelligence

2.1.2.1 Neural Networks

Neural networks are made up of a large number of neurons, massively connected. They make up the nervous system and the brain; the human brain may contain

10"11 neurons and 10"15 interconnections. Biological neural networks can be stated as groups of active neurons specialised in tasks such as: mathematical computation, positioning and memory (Simpson, 1990). (Edgar Sanchez et. al, 2006).

Advantages offered by neural networks

Due to their constitution and foundations, artificial neural networks exhibit a large number of features similar to those of the brain. For example, they are able to learn from experience, to generalise from previous cases to new cases, to abstract essential features from inputs that represent irrelevant information, and so on. This offers numerous advantages and this type of technology is being applied in many areas. The advantages include:

- Adaptive Learning. The ability to learn to perform tasks based on training or initial experience.
- Self-organisation. A neural network can create its own organisation or representation of the information it receives through a learning stage.
- Fault tolerance. Partial destruction of a network leads to a degradation of its structure, however, some network capabilities can be retained, even if it suffers extensive damage.
- Real-time operation. Neural computations can be performed in parallel; special hardware machines are designed and built to achieve this capability.
- Easy insertion into existing technology. Specialised chips can be obtained for neural networks to enhance their capability for certain tasks. This will facilitate modular integration into existing systems.

Adaptive learning.

The capacity for adaptive learning is one of the most attractive features of neural networks. That is, they learn to perform certain tasks by training with illustrative examples. Since neural networks can learn to differentiate patterns through examples and training, there is no need to develop a priori models and no need to specify probability distribution functions.

Neural networks are self-adaptive dynamic systems. They are adaptive due to the self-adjusting capacity of the process elements (neurons) that make up the system. They are dynamic, as they are capable of constantly changing to adapt to new conditions. In the learning process, the weighted links of the neurons are adjusted in such a way as to obtain certain specific results. A neural network does not need an algorithm to solve a problem, as it can generate its own distribution of link weights by learning. There are also networks that continue to learn throughout their lifetime, after their training period is completed. The function of the designer is only to obtain the appropriate architecture. It is not the designer's problem how the network will learn to discriminate. However, it is necessary for the designer to develop a good learning algorithm that will provide the network with the ability to discriminate, through pattern training.

Self-organisation.

Neural networks use their adaptive learning capability to self-organise the information they receive during learning and/or operation. While learning is the modification of each procedural element, self-organisation is the modification of the entire neural network to accomplish a specific goal. When neural networks are used to recognise certain kinds of patterns, they self-organise the information used. For example, the network, called backpropagation, will create its own characteristic representation, by which it can recognise certain patterns. This self-organisation results in generalisation: the ability of neural networks to respond appropriately when presented with data or situations to which it has not previously been exposed. The system can generalise the input to obtain a response. This feature is very important when solving problems in which the input information is not very clear; it also allows the system to provide a solution, even when the input information is incompletely specified.

Fault tolerance.

Neural networks were the first computational methods with the inherent capability of fault tolerance. Compared to traditional computational systems, which lose their functionality when they suffer a small memory error, in neural networks, if a failure occurs in a not very large number of neurons and although the behaviour of the system is influenced, it does not suffer a sudden drop. There are two distinct aspects to fault tolerance:

Networks can learn to recognise noisy, distorted or incomplete patterns. This is fault tolerance with respect to data.

- Networks can continue to function (with some degradation) even if part of the network is destroyed.

The reason neural networks are fault-tolerant is that they have their information distributed across the connections between neurons, and there is a degree of redundancy in this type of storage. Most algorithmic computers and data retrieval systems store each piece of information in a single, localised space. Therefore, most of the interconnections between network nodes will have their values based on the stimuli received, and an output pattern will be generated that represents the stored information.

Real-time operation.

One of the highest priorities in almost all application areas is the need to process data very quickly. Neural networks are well suited to this due to their parallel implementation. For most networks to operate in a real-time environment, the need for changing connection weights or training is minimal.

Easy insertion into the existing technology.

An individual network can be trained to perform a single, well-defined task (complex tasks, which make multiple pattern selections, will require interconnected network systems). With existing computational tools (non-PC type), a network can be quickly trained, tested, verified and transferred to a low-cost hardware implementation. Therefore, there are no difficulties for the insertion of neural networks in specific applications, e.g. control, within existing systems. In this way, the neural network can be used to improve systems incrementally and each step can be evaluated before undertaking further development.

The disadvantages of neural networks are:

Neural networks must be trained for each problem. In addition, multiple tests are necessary to determine the appropriate architecture. Training is time-consuming and can consume several CPU hours. Because the networks are trained rather than programmed, they require many parameters to be defined before the methodology can be applied. For example, the most appropriate architecture, the number of hidden layers, the number of nodes per layer, the interconnections, the transformation function, etc. must be decided. Conventional statistical techniques, however, only require the extraction and normalisation of a sample of data. It is an erroneous argument to argue that the development time for neural network-based models is shorter than the time needed to develop, for example, a multiple regression-based scoreboard. Studies where shorter development time has been found have not taken into account the data preparation required by a neural network.

Neural networks present a complex aspect for an outside observer wishing to make changes. To add new knowledge, it is necessary to change the interactions between many units so that their unified effect synthesises that knowledge. For a problem of significant size it is impossible to do this manually, so a network with distributed representation must employ some learning scheme.

2.1.2.2 Fuzzy Logics

Fuzzy logic, contrary to what its name indicates, is an alternative rationale to classical logic that aims to qualify imprecise information, i.e. with a certain degree of vagueness, emphasising that the fuzzy ones are the variables, which represent this information, and not the model as such. Fuzzy logic is a multivalued logic that allows to represent uncertainty and vagueness mathematically, providing formal tools for its treatment. As stated by (Zadeh, 1973), "As complexity increases, precise statements lose their meaning and useful statements lose their precision". Which can be summarised as "the trees don't let you see the forest". Basically any problem in the world can be solved as given a set of output variables (output space).

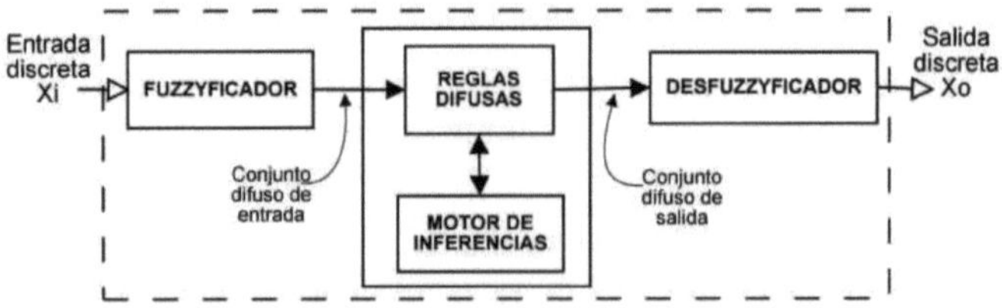

Fig.1. Introduction to fuzzy logic

The term fuzzy logic was first used in 1974. It is now used in a broad sense, grouping together fuzzy set theory, if-then rules, fuzzy arithmetic, quantifiers, etc. One of the aims of fuzzy logic is to provide formal support for reasoning in natural language, which is characterised by approximate reasoning using imprecise premises as a tool for formulating knowledge. Fuzzy logic was born, then, as the logic of approximate reasoning, and in that sense, it could be considered an extension of multivalued logic.

The concept of approximate reasoning can be interpreted as the process of drawing imprecise conclusions from imprecise premises. Zadeh introduced the theory of approximate reasoning and many other authors have made important contributions to this field. In natural language, objects or situations are described in imprecise terms: big, young and shy. Reasoning based on these terms cannot be exact, as they usually represent subjective impressions, perhaps probable, but not exact. Because of this, fuzzy set theory is more suitable than classical logic for representing human knowledge, as it allows phenomena and observations to have more than two logical states.

Fuzzy control is a methodology for the construction of control systems in which the correspondences between the actual valued inputs and the output parameters are represented by fuzzy rules. Fuzzy control has been very successful in commercial products such as automatic transmissions, video cameras and electric shavers. The main advantage of a fuzzy logic based control system is that it delivers outputs in a fast and accurate manner, thus reducing the fundamental state transitions in the physical environment it controls.

Fuzzy logic is a methodology that provides a simple way to obtain a conclusion from imprecise, noisy or incomplete input information. In general, it mimics how a person makes decisions based on information with the aforementioned characteristics. One of the advantages of fuzzy logic is the possibility of implementing fuzzy logic based systems in hardware, software or a combination of both.

Fuzzy logic is an artificial intelligence technique that allows working with information with a high degree of imprecision, in this it differs from conventional

logic that works with well-defined and precise information.
Man, in the quest for precision, attempted to fit the real world into rigid and static mathematical models, such as classical binary logic. When Aristotle and his predecessors devised their theories of logic and mathematics, they proposed the Law of the Excluded Centre which states that every matter must be either true or false. The grass is either green or not green, clearly it cannot be both green and not green.

2.2 Background.

The traffic light is also known as a traffic "controller", as it is responsible for managing the road infrastructure, keeping both vehicles and pedestrians safe. The first traffic light was installed in London and arose from the need to manage the roads safely; but due to the lack of technology at the time, the traffic light required a traffic police officer to control it at all times, which meant fully manual control (Aleman et al., 2014).
Nowadays, given the development of digital systems, information technology and telecommunications, the previous traffic light model is no longer remembered. We now have fully automatic systems that manage traffic by means of time scheduling; hence the need to precisely assign the time that the green light must remain on to allow vehicles to pass. Many cities around the world have completely discarded the use of fixed-time traffic signals, where the traffic engineer rigorously calculates the green light time based on statistical data obtained by counting vehicles at the intersection under study. The traffic engineer is in charge of planning each of the traffic light timings and phases, as well as pedestrian crossings (Aleman et al., 2014).
The fixed-time traffic light has been rapidly displaced by traffic lights with a coordinated architecture, where a "master" controller provides a time base that serves as a reference for a network of traffic lights. Generally, this network of traffic lights handles the same traffic density; their main advantage is that, because they are coordinated, they allow a queue to be quickly dissipated by assigning green lights along an entire route, allowing the driver to travel at a constant speed, an effect known as a "green wave". Coordinated systems, making use of the green wave, can give preference to police cars, ambulances, etc. to allow such vehicles to quickly exit a traffic jam (Aleman et al., 2014).
Another advance in traffic signal systems is the implementation of actuated traffic lights; these are used when there is a main road that is congested, and many secondary roads that handle little traffic flow. The implementation of these traffic signals consists of the installation of vehicle detectors on the secondary roads that inform the controller of the presence of vehicles; the controller maintains the green light on the main road as long as there are no vehicles on the secondary

roads; otherwise, the controller will assign the passage to the vehicles on the secondary roads and will maintain the green light as long as the flow of vehicles is constant and their maximum passage time has not expired (Aleman et al., 2014). Modelling and simulation of transport systems is widely used due to the high costs and potential safety issues associated with their prototyping in real traffic. Modelling and simulation play a key role in understanding real systems through abstraction and evaluating new technologies. In this work, a simulation-based analysis was performed to evaluate the potential benefits of applying pattern recognition in Intelligent Transportation Systems (ITS).

Moreover, in (Ibrahim, H., 2015) integrated transport models and a simulation software called AIMSUN are used, which is used to develop a model for a given road route. Also shown is the development of a series of plug-in software modules to implement pattern recognition, route guidance, and message communication functions. For the above, the authors used the developed model in two congestion scenarios, named reactive and proactive. The simulation results were promising and showed that the integration of pattern recognition into transportation systems improves mobility, reduces travel time, and reduces congestion resolution time. In particular, a 5% to 30% reduction in average travel time and an 8% to 41% reduction in average congestion resolution time were obtained.

As can be seen, the efforts made in the field of computing to support decisions in traffic light control is a task arduously studied in different parts of the world, which opens up the possibilities of solving different challenges observed over the last 30 years in terms of vehicular mobility and urban interconnection.

For example, in the early 1970s, the Autofahrer Leit and Information System (Autofahrer Leit and Information System) was developed in Germany, which is a dynamic system that proposes a route based on road conditions. The aim of this contribution was to solve common traffic problems such as traffic jams or to serve as the first intelligent assistance for drivers, just like the technology we use today to go to new places.

On the other hand, a joint work carried out between different governments, universities and private companies in 19 European cities established the PROMETHEUS project, where several technologies were tested during the period 1987 to 1994. These tests already had indications of the use of computational techniques or mathematical models to support the calculation of the timing of the lights, seeking to improve the service levels of the roads.

Now, another type of contribution that goes beyond road management systems were the first signs of intelligent cars, of which a first contribution can be found in the VaMoRs proposal, which was a project presented in Munich in 1991, as a prototype that used two television cameras with the objective of following a line.

In the same vein of autonomous car contributions, Vitall is a proposal that in a 1994 effort by Daimler-Benz designed a vehicle that used 10 cameras and 60 processors to keep the car on a line as it moved, maintaining a safe distance from the car in front of it.
Later, with the aim of generating instruments to support drivers on the roads, the proposal called DRIVE (Dedicated Road Infrastructure for Vehicle Safety) was born as a programme dedicated to the development and testing of communication systems for driving assistance tasks.
As all these contributions were revolutionising the way to face the traffic problem, but without a regulation or standardisation, the [2]ERTICO project (European Road Transport Telematic Implementation Coordination Organisation) was born in Brussels, Belgium in 1992, whose purpose was to provide support to the improvement and implementation of European transport projects, defining standards that are still in force today.
In 1970, Rosen et al (1970) proposed ERGS (Electronic Route Guidance System), the first effort recognised by the transport community and worked by the USA from the 1960's until the early 1970's which was basically an electronic navigation system. This system uses a two-way roadway to communicate to vehicles an ideal route.
In 2000 the MOBILITY 2000 study group laid the groundwork for the formation of the IVHS America (Intelligent Vehicle Highway System) which is a public and private forum to consolidate ITS interests and promote cooperation with other countries.
The AHS (Automated Highway System) was led by the NAHSC (National Automated Highway System Consortium) formed by the US Department of Transportation, General Motors, the University of California and other institutions. Under this project several fully autonomous vehicles were demonstrated on California roads.
Finally, by 1996, Japan launched the ITS (Intelligent Transportation System) initiative to catch up with the new methodological approach proposed by the United States of America.
In the early 1970's, the Comprehensive Automobile Traffic Control System (CATCS) was developed in Japan, which is a dynamic system that proposes a route based on road conditions. In the early 80's, Japan set up several projects under the RACS (Road Automobile Communication System) platform.
The Advanced Mobile Traffic Information and Communication Systems AMTICS project was founded by the Japanese Government in the late 1980s.
The ARTS Advanced Road Transportation System (ARTS) demonstrated the

[2] https://ertico.com

efforts made by the Ministry of Construction to present a system capable of integrating vehicle issues and road conditions to promote vehicle safety technology.

In the early 1990s, the National Police Agency of Japan brought together RACS and AMTICS to give way to VICS (Vehicle Information and Communication System), which provides the conditions of the route traced by the user using the coordinates of the car in question.

In 1991 the ASV (Advanced Safety Vehicles) project was developed which implemented and promoted the use of various electronic and computer technologies to raise the safety level of vehicles.

Finally, by 1996, Japan launched the ITS (Intelligent Transportation System) initiative to catch up with the new methodological approach proposed by the United States of America.

All these revolutionary road management philosophies introduced the metaphor of applying technologies, techniques or models from other disciplines in order to achieve the consolidation of tangible proposals that would provide solutions to the different problems experienced on the roads of cities around the world. In this way, some direct contributions to the state of the art of intelligent transport systems, which arise from basic research, are presented below.

To begin with we can see in Chen et al., (2005) that the authors propose to develop a model of an adaptive and cooperative agent for decentralised traffic signal control. In this work each traffic light has the ability to process traffic related information and interact with the rest of the traffic lights using this information. The authors show promising experimental results using this philosophy.

Later, Chen et al. (2006) establishes a new structure for traffic lights to operate individually and collectively by proposing a hybrid control architecture that bases the interaction of its entities (i.e., the traffic lights) on the FIPA Standard. Unlike previous work, there is now an entity above the traffic lights which is in charge of monitoring what each traffic light does and, if necessary, it enters into operation in conjunction with the traffic lights to support them in their decision making.

Using the FIPA interaction standard, the work presented in Garcia-Serrano et al., (2003) was able to design traffic agents in cities so that each intelligent entity could recommend to the traffic light what actions it should perform (i.e., modifications to the traffic light cycles). To achieve this, the agents used knowledge-based information about the roads.

On the other hand, Hernandez et al., (2002) presents a study and comparison of an agent-based architecture that can be centralised and decentralised for traffic management in urban networks. This hybrid control proposal aims to show the advantages of each model where first, the centralised control avoids generating redundancy in the management of information and decisions and on the other hand the decentralised model exploits the individual deductive and generative capacity of each intelligent agent.

In the reference Liu et al., (2005) the authors propose a framework by simulating a multi-agent system to evaluate demand-side usage in an urban bus system. In this proposal, each agent in the system is in charge of managing the locations where buses should make their passenger stops. The main idea in this proposal is that the bus will decrease its stops and therefore increase the level of service offered to passengers.

Wang (2005) and Wang (2008) describe both hardware and software implementations and their scope of development in an urban traffic controller which integrates the concept of mobile agents.

Finally, an interesting contribution can be found in Zhanh et al., (2004) which introduces an architecture dedicated to the management and processing of distributed traffic data.

State of the Art

Evaluating the characteristics that constitute the daily operation of a road (i.e., users, schedules, directions, etc.) is a complex task since it involves factors that are as diverse as they are different and mutually exclusive, which makes it very difficult to generate a standardised understanding of one road to another or of a street in a given city; and this is to a large extent the challenge that has kept the task described above in force.

Derived from this, in order to generate a proposal that attempts to offer an adequate service based on the characteristics of a given road, aspects such as the quality of the information to be used for decision making, the operational structure that the intelligence model intends and the scope of the situations that it intends to address must be considered. Therefore, it is necessary to evaluate previously used technology and detect the applied approach.

In fact, most researchers addressing the problem of road congestion have conducted studies in cities so that the proposal is modelled directly on real data, demonstrating promising results through the use of advanced computational techniques. In this sense, some of the research that has been used as a motivation and reflection guide for this proposal is presented below.

In order to collect these articles, a literature review was carried out following the methodology presented below:

1. SCOPUS databases are used because the Universidad Autonoma de Tamaulipas currently provides us with free access to its contents.
2. In the search process it was decided to consider only those articles that had a scope at different stages of the knowledge frontier.
3. The following keywords were used for the queries:
    - Intelligent Transport Systems.
    - Urban Traffic Optimization.
    - Autonomous Traffic Lights Control.
    - Intelligent Systems on Traffic.
    - Traffic Lights Optimization.

By carrying out the process described above, it was possible to develop a literature review of 200 articles, which are detailed below. It is important to note that for the scope of this research, some articles are only mentioned, beyond showing an explicit explanation of them.

For example, for almost 20 years in Roozemond (2001) the authors argue that the usability and efficiency of a system to manage or control traffic depends to a large extent on its ability to react to patterns and permutations of actions on the road. Furthermore, it is pointed out that there are factors that can negatively impact on the performance of systems, such as changes in traffic flow, accidents, demands, etc. Based on this, the paper introduces an intelligent agent-based urban intersection control system, which reacts to traffic changes in the environment and adapts to them in order to change traffic rules. The main conclusion of this research is that the proactive and reactive nature of intelligent agents is fully adequate to solve the challenges of such a relevant application area as traffic control, being applicable for any kind of traffic (i.e., air, land, sea).

In the same sense of the use of intelligent agents, in Garcia-Serrano *et al.,* (2003) the authors indicate that when trying to modulate urban traffic, the main challenge is to identify a technology that is capable of being as dynamic as the problem itself, being in the words of Roozemond (2001) ***"agents one of the best alternatives for urban traffic".*** However, a community of agents (i.e., a multi-agent system) has its own problems to solve, with communication and interaction being one of the most relevant at the level of group operation, and for the traffic problem, this is of great importance. Therefore, the authors propose to use the FIPA protocol to design a traffic agent that is able to make knowledge-based recommendations and interact with others like it managing other roads. The results obtained allow since then to appreciate that this is one of the most suitable ways to continue working on intelligent transport systems.

Being multi-agent systems one of the paradigms that began to emerge as a suitable alternative to operate intelligent traffic control systems, in Zhanh *et al.* (2004) the authors describe the development of a multi-agent architecture for distributed traffic data processing and management. In this contribution to the state of the art, we can see how each agent is able to obtain, study and evaluate the information of the road it is managing and uses this information to exchange it with the rest of the entities in the system. The results presented at the simulation level allow us to appreciate that multi-agent systems are efficient and effective in solving urban traffic problems just as a group of human officers would do, considering that the uncertainty of information capture is a challenge that was not addressed.

Taking into consideration that the area of intelligent transport systems encompasses more problems than urban traffic, in the reference Chen *et al.*

An adaptive and cooperative agent model for traffic light control is proposed to achieve decentralised traffic light control.

(Katwijk *et al.,* 2005) Develops a test environment to enable the design of multi-agent systems to perform experiments with different strategies to examine the performance of intelligent traffic systems.

(Liu *et al.,* 2005) proposes a multi-agent simulation framework to evaluate the usability demand of a bus system.

(Wang, 2005)

(Wang, 2008) Describes the software and hardware implementation of the integration of the mobile agent concept in the field of urban traffic control.

(Chen *et al.*, 2006)

(Mobile) Proposal using the FIPA communication standard to realise a hybrid control architecture.

(Perez *et al.*, 2010) proposes a new infrastructure for communication and intelligent speed control of vehicles, which uses RFID identification for the recognition of traffic signals by means of the sensor based on the Hall effect.

(Saeed *et al.*, 2011) An application of fuzzy logic for autonomous traffic light control based on a multi-agent approach using wireless sensors to solve problems such as congestion, accidents, speed and traffic irregularities is shown.

Zade and Dandekar, 2011) introduce an intelligent traffic light controller using FPGA to realise a neuro-fuzzy system. This proposal is able to make decisions to reduce delay times at an intersection.

(Huang *et al.*, 2012) A reservation-based approach at road intersections is presented. This approach proposes the development of a new test simulator, which integrates an emission analyser.

(Kulkarni, 2015) Uses a back propagation neural network for decision making on the timing of clearance, landing and obtaining the average air traffic.

(Z. Zhang et al., 2017) attempt to develop a hybrid deep learning framework that fuses convolutional neural networks and long-term memory neural networks to predict travel time on urban highways.

(Zhang and Kabuka, 2017) Combines the recurrent neural network and the closed recurrent unit to predict traffic flow taking into account weather conditions.

Khadilkar, 2018) Defines route assignments, arrival and departure times for all trains on the line, using a reinforcement learning approach.

(David & Guillermo, 2018) Notifies pilots to take appropriate action for situations they encounter using machine learning, gathering information about the aircraft, flight path, weather information, terrain, and information about the pilot's current state.

Methodology

4.1 Identification of the Main Study Variables in an Urban Road.

When a land vehicle is in circulation it can be for different reasons, being classified in three simple ways:

i) usopersonal .

ii) use use.

iii) service use.

Let us understand then that a vehicle for personal use is any vehicle that is used by its owner (in most cases) and whose main purpose is to make a trip, usually short, to a destination that in most cases is usually in a specific routine. That is to say, a person uses his car to go to work, to go to the supermarket (almost always to the one he prefers), to carry out some leisure activity (which are usually the same during certain stages of a person's life) or very sporadically to some unfrequented destination.

From another perspective, a commercial vehicle is understood to be one that has a commercial purpose and whose routes or destinations will depend largely on the type of service the vehicle provides. For example, a beverage delivery truck has a route and times assigned by its company, which must be respected. However, a parcel delivery truck will always have different routes and destinations from one day to the next. In this sense, this type of vehicle use is more challenging to analyse than personal use.

Finally, there are service vehicles which are used to serve customers on the road. Taxis, trucks, ambulances, fire brigades and police vehicles make journeys with different routes, destinations and/or characteristics. Taxis and trucks normally have a defined route but these vehicles develop a totally unpredictable behaviour as their stops and starts are a function of the service dynamics required by the customers, and this really applies to ambulances, police and fire brigades as well. In this sense, identifying which information will be adequate to make decisions to achieve adequate levels of service for a given road is a very broad challenge that needs to be evaluated from different perspectives. Throughout the literature one can find different appreciations of this challenge, where authors go back and forth using different data. Characteristically they all achieve promising and replicable results, which is a strong indication that the information to be used for decision making in traffic control should not be taken lightly and can be adjusted to the characteristics and scope of the research work.

Thus, after an extensive review of the literature, this research work defines the use of 8 variables which are grouped into 3 classes as presented in Table 4.

Table 4. Classes and Variables for Roads.

Classes	Variables
Intersection Restrictions	**Green Light Time** **Number of lanes on road**
User Qualities	**Type of Car Number of Cars on the Road Average Speed**
Conditions of the Day	**Time of Day Type of Day Period of Year**

For the sake of clarity, let us detail the classes as follows:
Intersection Constraints are those values that pertain to an intersection and that are defined by the conceptualisation that the road manager defines of it. The green light time is the main value of a traffic light phase. If the green light time of the traffic lights at a given intersection is known, the rest of the phase values can be defined. Obviously it is also necessary to know the operation model of the traffic lights because sometimes when there are more than two traffic lights there is the possibility that 2 of them are synchronised to have the same behaviour, on the other hand, there are intersections where each traffic light has its own green light and the rest are on red light. The amber or yellow light (as we colloquially call it) has a standardised duration in Mexico of 3 seconds.
Another variable that is defined by the city administration is the number of lanes that can be used on each road. There are cities in Mexico where you can drive in three lanes, but the administrators allow one lane to be used for parking. Other cities do not allow this in order to give more flow to cars. This dilemma belongs and will belong to each administrator. The confusing thing is that this can change every time the administration of a city changes and this affects the road users.
User qualities refer to those variables that depend entirely on the decisions a driver makes while making a journey in a motor vehicle. On the one hand and as we all know, unfortunately and in spite of traffic rules and regulations, there are drivers who behave in very dispersed ways while driving. On the other hand, it is impossible to control the usage and service times of a road as this is completely dependent on the needs of the individual driver. In spite of this, there are some works in the literature that promote the use of mathematical models to try to reproduce the behaviour of drivers and that have achieved promising results

Castan et al. There are 3 variables that are classified in the user qualities such as: i) type of car; ii) number of cars on the road and; iii) average speed.

i) Car type is the variable that allows to classify vehicles by their size, which is necessary to determine aspects such as the number of cars that can fit on a road or the estimated greenhouse gas count.

ii) The number of cars on a road is a necessary value for traffic control modelling as it is the predominant variable for calculating the traffic volume and the level of demand on that road.

iii) The average speed of a road allows modelling the behaviour of arrivals and departures of cars during the phases of each traffic light. It is necessary to know this data in order to properly estimate the behaviour of roads.

Daytime conditions must be known as they set the tone for the behaviour of a roadway. Each road and each intersection presents traffic situations differently and this behaviour is usually associated with and determined by the traffic schedule, the type of day it is and to a large extent by the season of the year. Normally the mathematical models used by traffic engineering experts do not take these variables into account due to the high level of uncertainty they introduce into the models.

The time of day needs to be known since the traffic conditions of a road are modulated based on the use of cars derived from the mobility needs of the inhabitants of a city. For example, based on the work and school schedules in Mexico, it is known that 7 am and 14 pm are extremely complicated schedules and on the other hand, at 5 am and 22 pm the opposite is estimated. In this way we can demonstrate the importance of knowing the time of day to be able to carry out an adequate control of a road system.

The type of day is another determining factor that must be taken into account when it comes to proper road management. As we all know, traffic is not the same at 7 am on Monday as it is on Saturday or Sunday. Or on Friday at 22 pm than on Sunday at the same time. Therefore, this variable must also be taken into account when implementing automatic learning systems to manage a road.

Finally, the time of year is another important factor to consider. In a normal school or work period, road conditions are usually very similar from one day to another, but this can vary if one of these days is a holiday (May 1st or November 2nd) or if the calendar dates are within a holiday period (Easter, Christmas, etc.). Thus, in order to manage urban traffic in an adequate way, this variable acquires an interesting relevance and makes that any intelligent vehicle control system must take it into account.

Table 5 presents each of the variables with the type of data it will handle and a brief justification of why the variable will be used in our proposal.

Table 5. Variables: type of data and justification.

Variable	Type of Datum	Justification
Green Light Time	Numerical Whole	Indicates the time that exists at the intersection for cars to exit.
Number of lanes on road	Numerical Whole	Explains the number of lanes that can be used on a given road.
Type of car	Descriptive	Presents each car sorted by size.
Number of cars on the road	Numerical Whole	Control the number of cars on each road when making decisions.
Average Speed	Numerical Not Whole	Displays the average arrival and departure speed of cars at an intersection.
Time of day	Numerical Not Whole	It allows to know the time when decisions will be taken.
Type of day	Descriptive	It allows the type of day on which decisions will be taken to be known.
Period of the Year	Descriptive	It allows to know the time of the year when decisions will be taken.

Specifically for each of these variables, some attributes were predefined to facilitate the interpretation of the information, particularly in those variables whose data type is descriptive. On the other hand, for the variables of type number (integer and non-integer) free data from 0 to an infinite value are expected. The variables defined above and their attributes are presented below.

Table 6. Numerical variables and the ranges of their attributes.

Variables	Lower Range	Top Rank
Green Light Time	25 seconds	45 seconds
Number of lanes on road	1	
Number of cars on the road	0	Infinite
Average Speed	20 km/h	80 km/h
Time of day	0:00 hours	23:59 hours

Table 7. Descriptive Variables and their attributes

Variable	Attributes
Type of car	Personal use
	Commercial use
	Service use
Type of day	Weekdays
	Weekend
Period of the Year	Normal
	Holiday
	Holidays

4.2. Standardisation of variables.

As can be seen in the literature, most, if not all, authors agree on the need to preprocess the information to facilitate the interpretation of machine learning techniques. Fortunately there is a wide range of methods that allow this task to be performed adequately, the situation is that there is no standardisation to affirm which is the best of them. This leads us to the challenge of implementing any of these and using tools that allow us to compare performances and find out which one might be the most suitable for the problem at hand.

Based on the above, 4 standardisation techniques are used in this research project, as follows:

4.2.1. Standard Standardisation.

This model is usually one of the most recurrent in the state of the art, as it allows working with large amounts of data and normalises the errors when the population parameters are known, as is the case in this study.

$$data = \frac{X - \mathrm{jU}}{\mathrm{a}} \quad (1)$$

In this sense, for the scope of this research it is proposed to use two granulates based on coarse grain and fine grain. It is important to define that due to the type of study some of the data cannot be preprocessed with this technique and will be classified based on the author's criteria.

4.2.3.1 Coarse Grain

In order to establish a more closed classification range, which generates a study focused on not allowing too much amplitude in the criteria for recommending products to customers, it is proposed to establish a granularity in four levels, based on the structure of the quartiles of statistical classification.

4.2.3.2 Thin Grain

Contrary to what is proposed in the coarse graining, the fine graining attempts to generate as many alternatives as possible so that the information can be classified and this generates a study more focused on the specific details of the analysis. Specifically, it is proposed to divide each of the quartiles of the coarse grain into three classes of decreasing level.

4.3. Behaviour of automatic learning techniques.

A machine learning technique used in the field of data mining and big data is a powerful analytical tool that allows to study trends and behaviour of information in an agile and efficient way. For this reason, this research proposes to study four techniques to determine which will be the most effective in recommending products to customers with different characteristics. In this way the techniques studied are:

- K nearest neighbours (KNN).
- K meneras (K means).
- Support Vector Machine (SVM).
- Decision trees.

4.3.1 Method Settings

Based on the methodology to be followed within the selected models, each of them requires some pre-study data to be established. It is important to mention that to validate all methods it is proposed to use 70% of the data for training and the rest for validation.

Specifically for models KNN and K means the following is proposed:

1. To calculate the distance between the item to be classified and the rest of the items in the dataset, we propose to implement 3 similarity measures in order to evaluate which of them is the most appropriate for the type of data we are working with. The selected measures are Cosine Similarity, Jaccard Index and Sorence-Dice.
2. For the precision tests, it is proposed to use the following values for K:

Table 8. K-values for Tests.

Test	K-value
1	
	5
5	

4.4. The Neuro-Fuzzy Model.

The decision model proposed to be presented in this dissertation seeks to address some of the challenges of today's roads by means of a neuro-fuzzy decision process. This involves the use of two advanced computational techniques such as fuzzy logic and neural networks. The intended scheme is shown in Fig. 1.

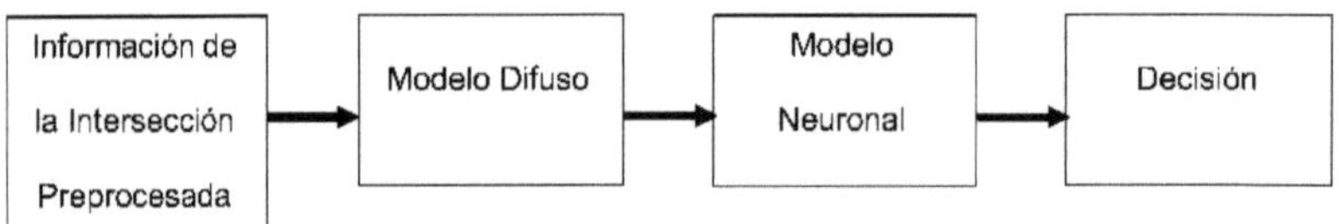

Fig. 1. Neuro-fuzzy diagram of the research proposal.

The information must be preprocessed to avoid noise at the time of modelling. The information is then analysed by the fuzzy model, which is responsible for defining a value for each of the three input neurons, appropriate to the behaviour patterns observed at the intersections studied in the vehicle gauging. A graphical representation of the process is shown in Fig. 2.

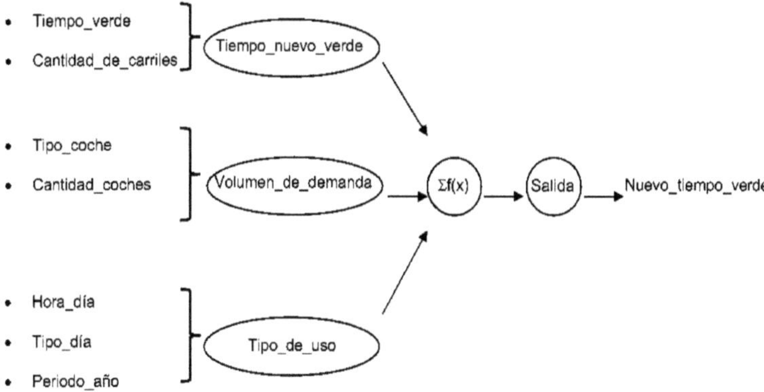

Fig. 2. Graphical representation of the system.

MatLab tools were used to generate the fuzzy model. A 3 neurons input layer. In fact, each model was realised with a Mamdani solution approach.

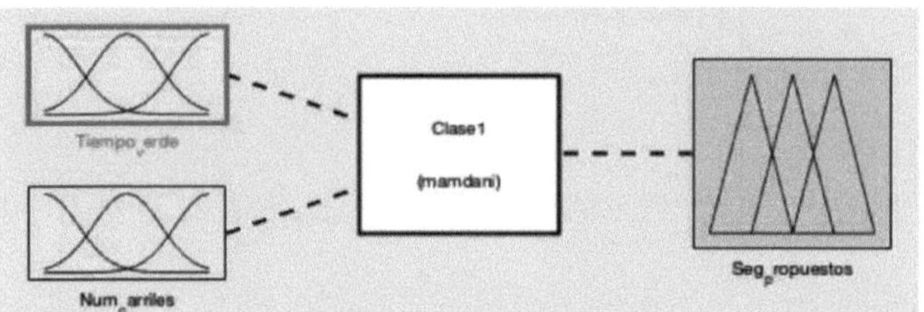

Fig. 3. Class 1 Fuzzy System.

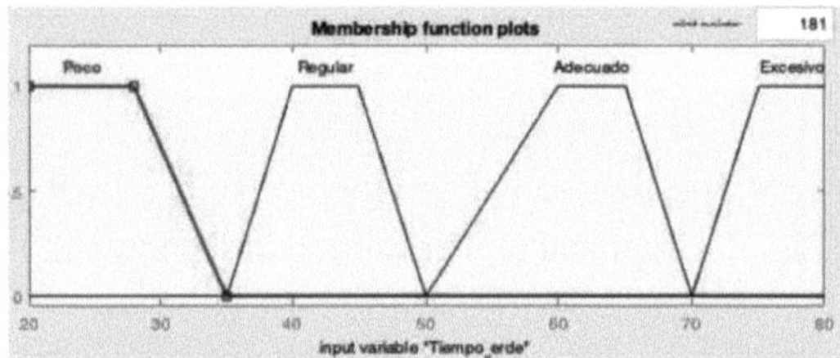

Fig. 4. Membership function of the Green_Time Variable.

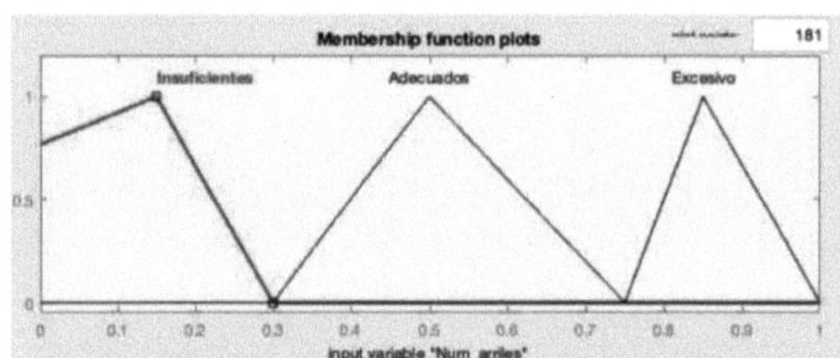

Fig. 5. Membership function of the Variable Number_of_rails.

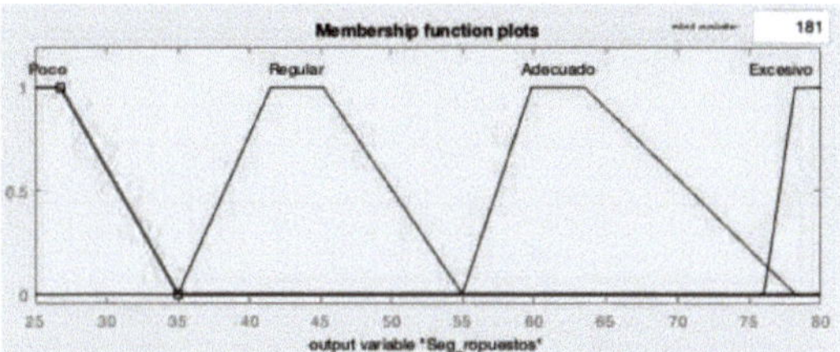

Fig. 6. Membership function of the Proposed_Seconds_Output.

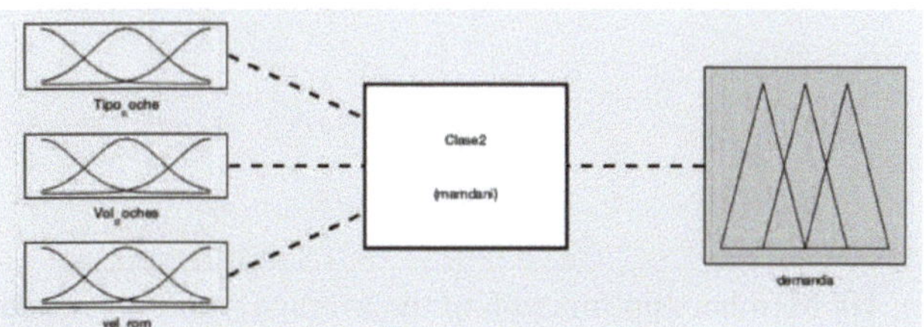

Fig. 7. Class 2 Fuzzy System.

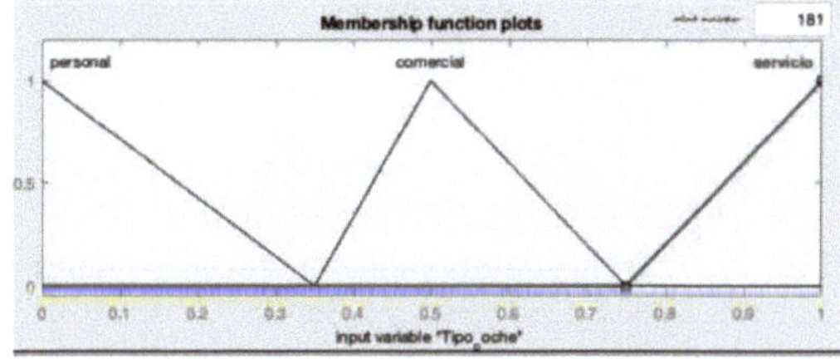

Fig. 8. Membership function of the Variable Car_type.

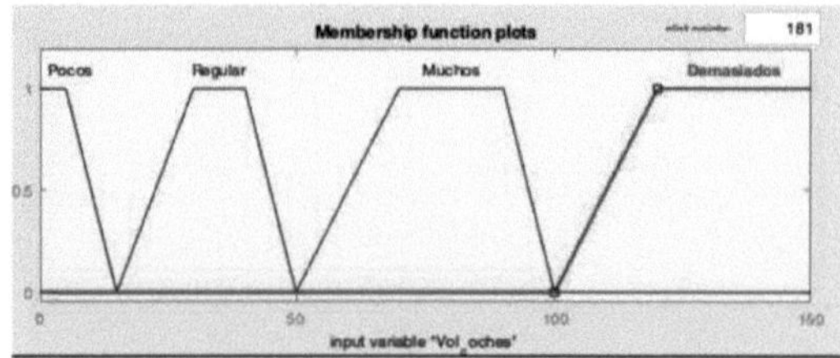

Fig. 9. Membership function of the car_volume variable.

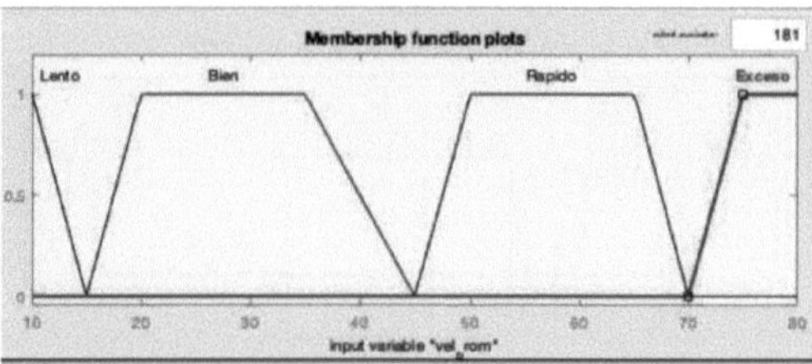

Fig. 10. Membership function of the average_velocity variable.

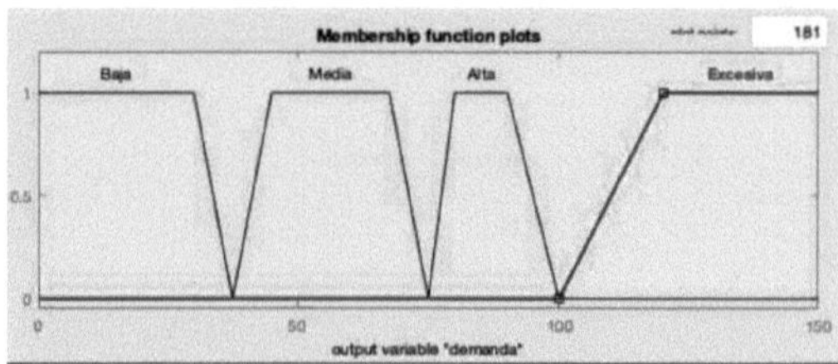

Fig. 11. Output membership function Demand.

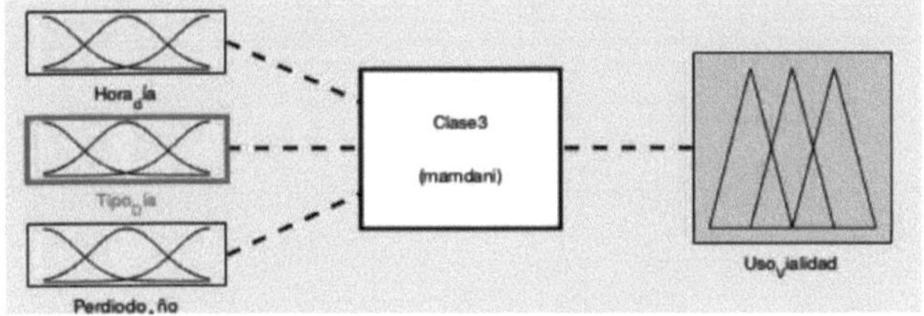

Fig. 12. Class 3 Fuzzy System.

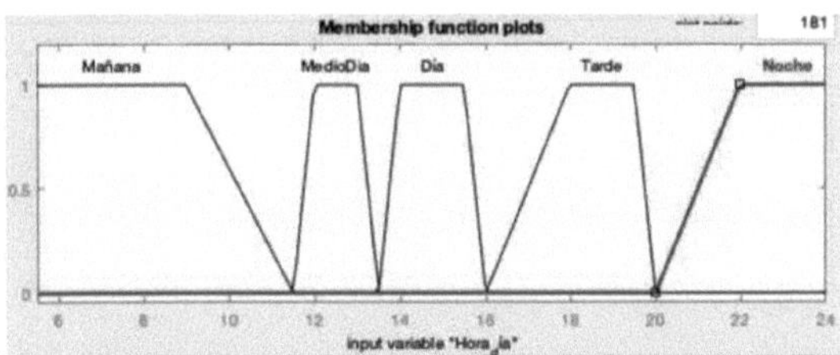

Fig. 13. Membership function of the Hour_Day_Variable.

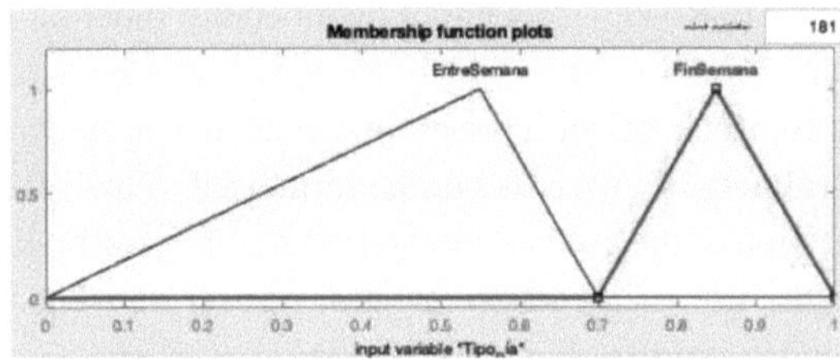

Fig. 14. Membership function of the Day_Type variable.

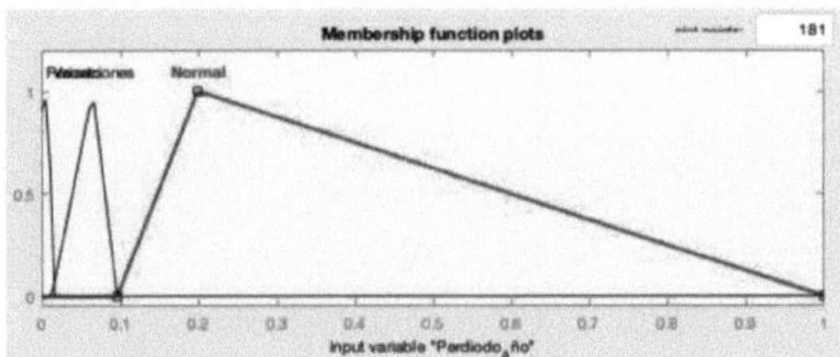

Fig. 15. Membership function of the Variable Period_Year.

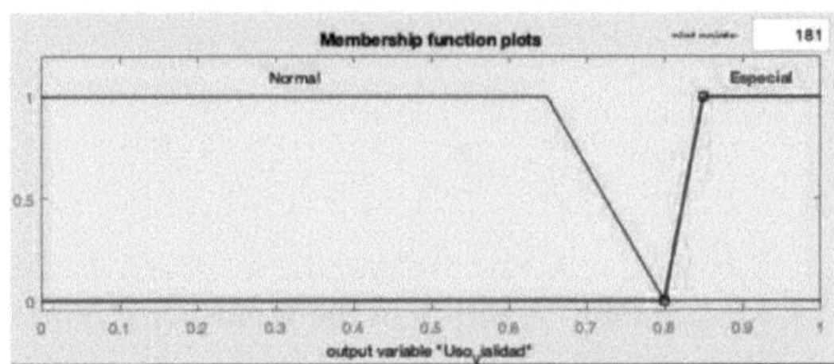

Fig. 16. Road_Use_Output membership function.

Having the data for each of the classes in which the 8 study variables were grouped, the neural network process can be initialised, which is responsible for determining a duration of the green light interval for each traffic light. To achieve this, we propose the use of a backpropagation neural network with a sigmoid activation function, as expressed in equation (3).

$$oW = \frac{1 A x}{1 + e^{x}} \quad (3)$$

Where e denotes the exponential constant and whose estimated value is the one that is close to
2.71828.

To establish the training of the neural network, 2,000 cases of the gauging observations were used. The complete model is run and the results obtained from the network are taken as a reference to calculate how many vehicles would leave the intersection with the set time by making a simple comparison against the green light time and the amount of cars that left in the real situation. Thus, Fig. 17 shows the results of the training phase. The neural network stabilises after training 1600 and reaches a stable performance of almost 80%.

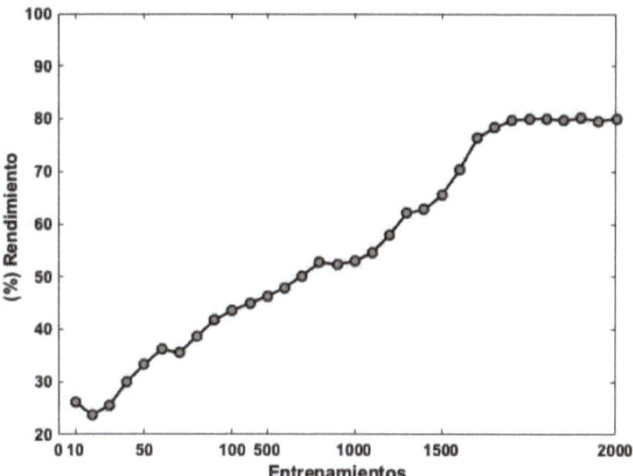

Fig. 17. Results of the Training Phase.

4.5. Design of experiments.

In order to measure the performance of the proposal made in this research work, a set of experiments is defined in which data is taken randomly from the gauges obtained, without chronological order and without taking into account which intersection the data belonged to. In this way it is intended to expose the system to a series of uncontrolled data to bring the tests as close to the highest level of uncertainty and openness as possible, as it happens in a real intersection.

The gauging was carried out at 3 different intersections over a period of 30 days with a duration of 10 hours. On average, the gauging shows that 250 readings were captured per day at each intersection. Calculating this and with the data in the database, the total number of gaugings is 22,500.

Specifically, 5 experiments were carried out. For each experiment a data from the gauging is taken and introduced into the proposed methodology. In fact, 70% of the data is used for a total of 15,750 experiments and it is intended to evaluate the level of efficiency of the system from the following points of view:

1. The performance of each experiment.
2. The stability of the decisions made by the proposed system.

4.6. Technical Feasibility Study of the Proposal.

This research involves performing large amounts of operations and calculations at the level of computer processing and using appropriate programming languages for its implementation. In this sense, from software engineering it can be distinguished how a computer system must be evaluated from the perspectives of performance, reliability, security, efficiency, portability, accessibility and information integrity, where such measures can be achieved on the basis of the selected programming language.

Therefore, it is important to select the language that meets the necessary requirements to give technical support to the research proposal. In order to study the above, the TIOBE classification of programming languages was considered. In this sense, Fig. 18. presents the TIOBE classification of programming languages, in its June 2020 version. TIOBE[3] is a company dedicated to the evaluation and monitoring of software quality. They measure software quality by applying a standard that is widely accepted by the worldwide community of developers. It is important to mention that they propose such a classification based on a generalised measurement that considers programming languages used for education, commercial applications, business administration, industrial processes, among others.

Jun 2020	Jun 2019	Change	Programming Language	Ratings	Change
1	2	↑	C	17.19%	+3.89%
2	1	↓	Java	16.10%	+1.10%
3	3		Python	8.36%	-0.16%
4	4		C++	5.95%	-1.43%
5	6	↑	C#	4.73%	+0.24%
6	5	↓	Visual Basic	4.69%	+0.07%
7	7		JavaScript	2.27%	-0.44%
8	8		PHP	2.26%	-0.30%
9	22	⇑	R	2.19%	+1.27%
10	9	↓	SQL	1.73%	-0.60%
11	11		Swift	1.46%	+0.04%
12	15	↑	Go	1.02%	-0.24%
13	13		Ruby	0.98%	-0.41%
14	10	⇓	Assembly language	0.97%	-0.51%
15	18	↑	MATLAB	0.90%	-0.16%
16	16		Perl	0.82%	-0.36%
17	20	↑	PL/SQL	0.74%	-0.19%
18	28	⇑	Scratch	0.73%	+0.20%
19	19		Classic Visual Basic	0.65%	-0.42%
20	38	⇑	Rust	0.64%	+0.38%

Figure 18. TIOBE classification.

Source. TIOBE Index

[3] https://www.tiobe.com/

On the other hand, Fig. 19 shows the classification of programming languages.

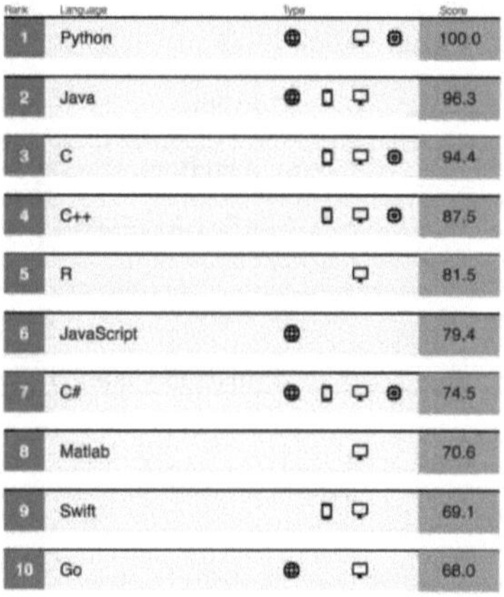

Rank	Language	Score
1	Python	100.0
2	Java	96.3
3	C	94.4
4	C++	87.5
5	R	81.5
6	JavaScript	79.4
7	C#	74.5
8	Matlab	70.6
9	Swift	69.1
10	Go	68.0

Fig. 19. IEEE Spectrum classification.

In this sense, for the development of the proposal it was decided to use the Matlab programming language, 15th in the TIOBE ranking and 8th in IEEE Spectrum. The selection of the language consisted of weighing the benefits of those languages that firstly facilitate the handling of large amounts of data organised indistinctly and in files with low memory space consumption, which can be carried out in a simple and user-friendly way in Matlab through the use of text files (txt extension); secondly, that allow the connection with other information processing tools such as cameras or embedded computing systems for possible future extrapolations of the proposal.

On the other hand, as mentioned at the end of the previous paragraph, for the scope of the tests carried out in this research it was necessary to use a tool specialised in information analysis. Specifically, WEKA[4] was used, which is a powerful platform for information analysis that uses automatic learning and data mining techniques written in JAVA and developed by the University of Waikato. In the same way that the software issues were analysed, it is necessary to evaluate the hardware characteristics. Thus, in Table 9 the technical specifications of the computer equipment used are as follows:

[4] https://www.cs.waikato.ac.nz/ml/weka/

Table 9. Computer Equipment Specifications.

Detail	Specification
Manufacturer	Apple.
Model	MacBook Pro 2018
Processor	Intel Core i7 CPU 3.20 GHz
Ram Memory	8 GB
Operating System	macOS Catalina Ver 10.15.5

Results

5.1 Evaluation of Similarity Measures.

Similarity measures are used to look for equality between data within a large set of information. For this research this is a very relevant process as it seeks to find the best classification model to determine the values of the 8 variables defined in the study. In this sense, the results obtained at the level of sensitivity at the moment of seeking to find the greatest amount of information so that later the classification methods can carry out the processes in a more adequate manner are presented below.

It is important to mention that, according to the analysis carried out in the state of the art, the aim of this research is to identify the similarity model that recovers the greatest number of vectors within a certain level of closeness, that is to say, that there is an equality of at least 85% between the search vector and the vectors located in the database.

For the tests, 1000 readings from different intersections were used, which were randomly selected, taking care that the data were not repeated. Each reading was converted into a vector. These vectors were compared against all other vectors in the database, including the vector itself, to count the number of similar cases retrieved by each similarity measure, where X is the component containing the elements of the search vector and Y is each of the vectors indexed in the database, each model is expressed as follows:

$$coseno(\bar{X},\bar{Y}) = \frac{\sum_{i=1}^{n}(x_i * y_i)}{\sqrt{\sum_{i=1}^{n}(x^2) * \sum_{i=1}^{n}(y^2)}} \qquad (4)$$

$$jaccard(\bar{X},\bar{Y}) = \frac{\sum_{i=1}^{n}(x_i * y_i)}{\sum_{i=1}^{n}(x^2) * \sum_{i=1}^{n}(y^2) - \sum_{i=1}^{n}(x_i * y_i)} \qquad (5)$$

$$dice(\bar{X},\bar{Y}) = \frac{2 * \sum_{i=1}^{n}(x_i * y_i)}{\sum_{i=1}^{n}(x^2) + \sum_{i=1}^{n}(y^2)} \qquad (6)$$

The results of this search are shown in Annex 1. To give an example, table 10 shows the results obtained by 5 search processes and at the end the average number of cases retrieved for each similarity measure.

Table 10. Example of Cases Recovered.

Test	Cosine	Jaccard	Says
1	580	95	170
50	813	287	346
	774		228
	941	190	363
850	946	81	164
Media	780		275

To corroborate the suitability of these data for the proposed study, a statistical study of the results was carried out using the Friedman model. Due to the nature of the data and the characteristics of the study, it was necessary to standardise the results using the normal standardisation formula. As can be seen in Figure 20, the results obtained from the Friedman statistic determine that it is accepted that there is no difference between the values of the 3 measures with the p-value being the p-value.

Holm / Hochberg / Holland / Rom / Finner / Li Table for $\alpha = 0.05$ (FRIEDMAN)

i	algorithm	$z = (R_0 - R_i)/SE$	p	Holm/Hochberg/Hommel	Holland
2	Coseno	1.697056274847703	0.08968602177036669	0.025	0.025320565519103666
1	Jaccard	0.21213203435596445	0.8320040285726364	0.05	0.050000000000000044

Bonferroni-Dunn's procedure rejects those hypotheses that have a p-value ≤ 0.025.

Holm's procedure rejects those hypotheses that have a p-value ≤ 0.025.

Hommel's procedure rejects those hypotheses that have a p-value ≤ 0.025.

Holland's procedure rejects those hypotheses that have a p-value $\leq 0.025320565519103666$.

Finner's procedure rejects those hypotheses that have a p-value $\leq 0.025320565519103666$.

Li's procedure rejects those hypotheses that have a p-value $\leq 0.008841893233019135$.

Fig. 20. Results of the Friedman statistic.

However, in a second test of Holm's fitted values, it was determined that there is a difference (albeit very small) with Cosine Similarity showing the best results.

Average Rankings of the algorithms (Friedman)

Algorithm	Ranking
Coseno	2.15
Jaccard	1.9400000000000015
Dice	1.9100000000000015

Friedman statistic (distributed according to chi-square with 2 degrees of freedom: 3.4200000000012665. P-value computed by Friedman Test: 0.18086579262213998.

Fig. 21. Results applying Holm adjustment.

Therefore, and as a preliminary conclusion, we can choose Cosine Similarity as the most stable model for the type of data being manipulated in this experiment. In this sense, for all studies where a similarity measure must be applied to determine closeness between the study values, the measure to be implemented is Cosine Similarity.

5.2 Implementation of Classification Techniques.

For each model, the confusion matrix and precision measures were obtained using each of the 4 normalisation techniques proposed in this research. In addition, the sensitivity, accuracy, specificity and F1 score results are presented for which the following equations were used. The results are then broken down by each technique.

$$Precisión = \frac{Verdaderos\ Positivos}{Verdaderos\ Positivos + Falsos\ Positivos} \quad (7)$$

$$Sencibilidad = \frac{Verdaderos\ Positivos}{Verdaderos\ Positivos + Falsos\ Negativos} \quad (8)$$

$$Exactitud = \frac{Verdaderos\ Positivos + Verdaderos\ Negativos}{Total\ de\ Casos} \quad (9)$$

$$Especificidad = \frac{Verdaderos\ Negativos}{Verdaderos\ Negativos + Falsos\ Positivos} \quad (10)$$

$$score\ F1 = \frac{2 * (Precisión * Sencibilidad)}{Precisión + Sencibilidad} \quad (11)$$

K		Model: Positive	Model: Negative
K = 3		Positive	Negative
Real Data	Positive	169	45
	Negative	109	77
K = 5		Positive	Negative
Real Data	Positive	186	
	Negative	107	
K = 7		Positive	Negative
Real Data	Positive		59
	Negative		
K = 9		Positive	Negative
Real Data	Positive		77
	Negative	106	

Table 14. Results for the KNN model in Standard Normalization.

K = 11		Model	
		Positive	Negative
Data Real	Positive	135	87
	Negative		59

Table 15. Measurements for the KNN model in Normalization Standard.

K-value	Precision	Sensitivity	Accuracy	Specificity	F1 Score
	0.79	0.61	0.71	0.29	0.69
5	0.85	0.63	0.73	0.24	0.73
	0.74	0.60	0.75	0.35	0.66
	0.66	0.58	0.75	0.42	0.62
	0.61	0.53	0.74	0.42	0.57

Table 16. Results for the KNN model in Uniform Normalisation.

K = 3		Model	
		Positive	Negative
Data Real	Positive		
	Negative		69
K = 5		Model	
		Positive	Negative
Data Real	Positive		46
	Negative	110	65
K = 7		Model	
		Positive	Negative

Data Real	Positive		
	Negative		

K = 9		Model	
		Positive	Negative
Data Real	Positive		
	Negative		

K = 11		Model	
		Positive	Negative
Data Real	Positive	148	71
	Negative		

Table 17. Measures for the KNN model in Normalization Uniform.

K-value	Precision	Sensitivity	Accuracy	Specificity	F1 Score
	0.75	0.58	0.72	0.32	0.65
5	0.80	0.62	0.75	0.29	0.70
	0.76	0.60	0.74	0.33	0.67
	0.71	0.55	0.71	0.33	0.62
	0.68	0.55	0.73	0.37	0.61

Table 18. Results for the KNN model in coarse granules.

K = 3		Model	
		Positive	Negative
Data Real	Positive		65
	Negative	109	
K = 5		Model	
		Positive	Negative
Data Real	Positive		69
	Negative	109	70
K = 7		Model	
		Positive	Negative
Data Real	Positive		59
	Negative		
K = 9		Model	
		Positive	Negative
Data Real	Positive		
	Negative	135	69
K = 11		Model	
		Positive	Negative
Data Real	Positive	135	65
	Negative		

Table 19. Measurements for the KNN model in Coarse Grain.

K-value	Precision	Sensitivity	Accuracy	Specificity	F1 Score
	0.70	0.59	0.73	0.37	0.64
5	0.69	0.58	0.74	0.39	0.63
	0.71	0.52	0.67	0.31	0.60
	0.71	0.51	0.65	0.30	0.59
	0.68	0.50	0.67	0.33	0.58

Table 20. Results for the KNN model in Thin Granules.

K = 3		Model	
		Positive	Negative
Data Real	Positive		63
	Negative	110	
K = 5		Model	
		Positive	Negative
Data Real	Positive		70
	Negative	108	69
K = 7		Model	
		Positive	Negative
Data Real	Positive		
	Negative	134	
K = 9		Model	
		Positive	Negative
Data Real	Positive		
	Negative	135	69
K = 11		Model	
		Positive	Negative
Data	Positive	135	65

Real	Negative		

Table 21. Measurements for the KNN model in Thin Granules.

K-value	Precision	Sensitivity	Accuracy	Specificity	F1 Score
	0.71	0.58	0.72	0.36	0.64
5	0.69	0.59	0.74	0.39	0.63
	0.72	0.52	0.67	0.30	0.60
	0.71	0.51	0.65	0.30	0.59
	0.68	0.50	0.67	0.33	0.58

Algorithm K means.

Table 22. Results for the K means model in Standard Normalization.

K = 3		Model	
		Positive	Negative
Data Real	Positive		48
	Negative	109	71

K = 5 Model

K = 5		Model	
		Positive	Negative
Data Real	Positive	183	
	Negative	108	75
K = 7		Model	
		Positive	Negative
Data Real	Positive	163	58
	Negative	114	65
K = 9		Model	
		Positive	Negative
Data	Positive	148	

Real	Negative	104	70
K = 11		Model	
		Positive	Negative
Data Real	Positive		86
	Negative		

Table 23. Measures for the K means model in Standard Normalization.

K-value	Precision	Sensitivity	Accuracy	Specificity	F1 Score
	0.78	0.61	0.73	0.31	0.69
5	0.84	0.63	0.72	0.24	0.72
	0.74	0.59	0.74	0.34	0.65
	0.70	0.65	0.59	0.75	0.43
	0.63	0.62	0.53	0.75	0.42

Table 24. Results for the K means model in Normalization Uniform.

K = 3		Model	
		Positive	Negative
Real Data	Positive		
	Negative		69

K = 5		Model	
		Positive	Negative
Real Data	Positive		46
	Negative	110	65

K = 7		Model	
		Positive	Negative
Real Data	Positive		
	Negative		

K = 9		Model	
		Positive	Negative
Real Data	Positive		
	Negative		

K = 11		Model	
		Positive	Negative
Real Data	Positive	148	71
	Negative		

Table 25. Measures for the K means model in Uniform Normalisation.

K-value	Precision	Sensitivity	Accuracy	Specificity	F1 Score
	0.75	0.58	0.72	0.32	0.65
5	0.80	0.62	0.75	0.29	0.70
	0.76	0.60	0.74	0.33	0.67
	0.71	0.55	0.71	0.33	0.62
	0.68	0.55	0.73	0.37	0.61

Table 26. Results for the K means model in coarse granules.

K = 3		Model	
		Positive	Negative
Data Real	Positive		65
	Negative	109	
K = 5		Model	
		Positive	Negative
Data Real	Positive		69
	Negative	109	70
K = 7		Model	
		Positive	Negative
Data Real	Positive		59
	Negative		
K = 9		Model	
		Positive	Negative
	Positive		
Data Real	Negative	135	69
K = 11		Model	
		Positive	Negative

Data Real	Positive	135	65
	Negative		

Table 27. Measurements for model K means in Coarse Grained.

K-value	Precision	Sensitivity	Accuracy	Specificity	F1 Score
	0.70	0.59	0.73	0.37	0.64
5	0.69	0.58	0.74	0.39	0.63
	0.71	0.52	0.67	0.31	0.60
	0.71	0.51	0.65	0.30	0.59
	0.68	0.50	0.67	0.33	0.58

Table 28. Results for the K means model in Thin Granules.

K = 3		Model	
		Positive	Negative
Data Real	Positive		63
	Negative	110	
K = 5		Model	
		Positive	Negative
Data Real	Positive		70
	Negative	108	69
K = 7		Model	
		Positive	Negative
Data Real	Positive		
	Negative	134	
K = 9		Model	
		Positive	Negative
Data Real	Positive		
	Negative	135	69

K = 11		Model	
		Positive	Negative
Data Real	Positive	135	65
	Negative		

Table 29. Measurements for the K means model in Thin Granules.

K-value	Precision	Sensitivity	Accuracy	Specificity	F1 Score
	0.71	0.58	0.72	0.36	0.64
5	0.69	0.59	0.74	0.39	0.63
	0.72	0.52	0.67	0.30	0.60
	0.71	0.51	0.65	0.30	0.59
	0.68	0.50	0.67	0.33	0.58

Support Vector Machine

Table 30. Results for the SVM model.

Norm. Standard		Model	
		Positive	Negative
Data Real	Positive		
	Negative	115	
Norm. Uniform		Model	
		Positive	Negative
Data Real	Positive		
	Negative	120	
Coarse Grain		Model	
		Positive	Negative
Data Real	Positive		
	Negative		
Thin Grain		Model	
		Positive	Negative
Data Real	Positive		
	Negative	120	

Table 31. Measures for the SVM model.

Norm	Precision	Sensitivity	Accuracy	Specificity	F1 Score
Standard	0.84	0.60	0.70	0.22	0.70
Uniform	0.68	0.54	0.69	0.35	0.60
Grain Thick	0.70	0.55	0.68	0.34	0.61
Grain Delgado	0.68	0.54	0.69	0.35	0.60

Decision trees

Table 32. Results for the Decision Trees model.

Norm. Standard		Model	
		Positive	Negative
Data Real	Positive	146	63
	Negative	114	77
Norm. Uniform		Model	
		Positive	Negative
Data Real	Positive	134	
	Negative	111	81
Coarse Grain		Model	
		Positive	Negative
Data Real	Positive		
	Negative	111	
Thin Grain		Model	
		Positive	Negative
Data Real	Positive		80
	Negative	104	

Table 33. Measures for the Decision Trees model.

Norm	Precision	Sensitivity	Accuracy	Specificity	F1 Score
Standard	0.70	0.56	0.70	0.36	0.62
Uniform	0.64	0.55	0.69	0.40	0.59
Grain Thick	0.65	0.55	0.70	0.40	0.60
Grain Delgado	0.62	0.56	0.71	0.43	0.59

5.3 Validation Experiments.

This section is dedicated to show the results obtained from the experiments phase where the proposed methodology is tested against real data obtained in the gauging. As mentioned in the previous chapter, for each experiment a random data is taken from the database and thus we have 15,750 tests for each of the 5 experiments carried out.

The results are presented graphically below.

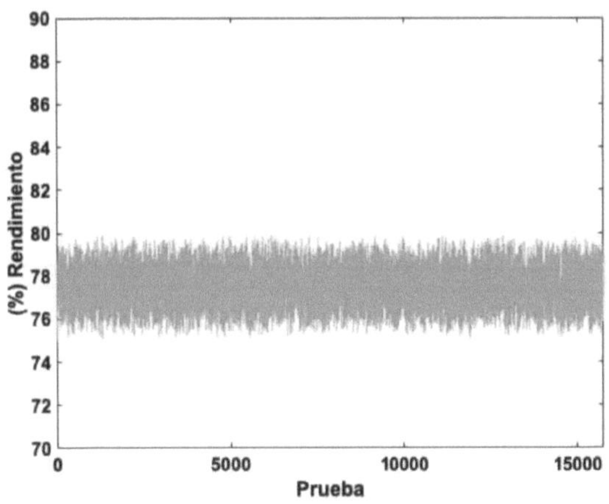

Fig. 22. Yield obtained from Experiment 1.

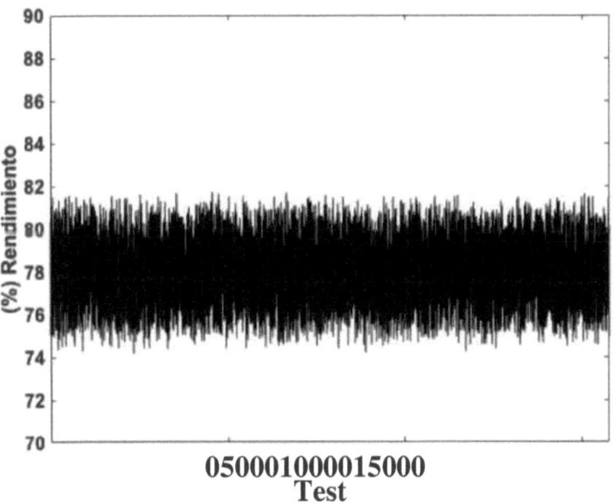

Fig. 23. Yield obtained from Experiment 2.

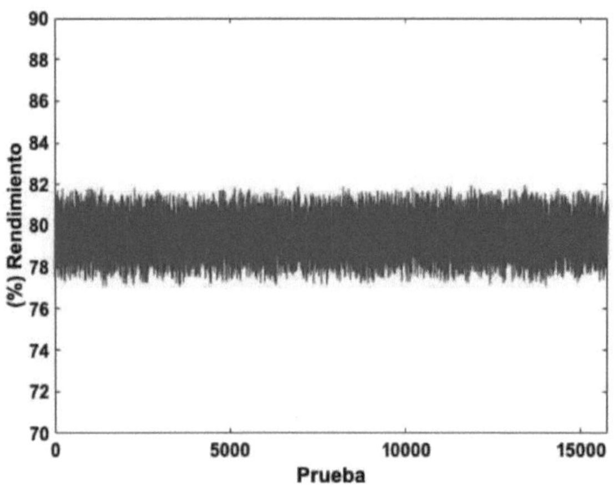

Fig. 24. Yield obtained from Experiment 3.

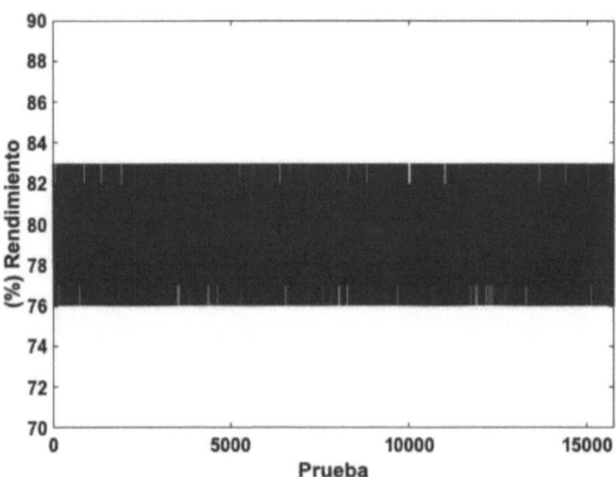

Fig. 25. Yield obtained from Experiment 4.

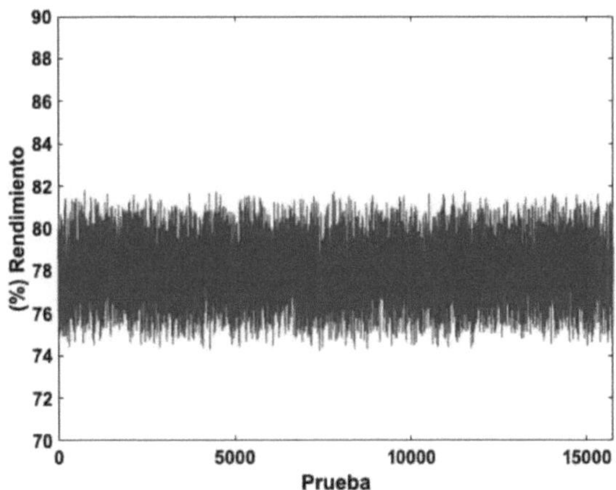

Fig. 26. Yield obtained from Experiment 5.

Conclusions and Future Work

6.1 Research Findings

Derived from the achievements obtained in this research and based on the objectives set out at the beginning of this research, the following conclusions are drawn:

It was possible to identify the main characteristics of the road, the type of service provided and the users in order to create a grouping model. This information made it possible to carry out some studies by implementing different techniques at different times during the development of the research. It can be said that the data obtained made it possible to carry out this research in a timely manner. It is worth mentioning that the surveys were carried out with the maximum effort to obtain the information as close to reality as possible, however, due to the great commitment shown by the support staff, the veracity of the information is not the responsibility of the authors.

Once access to the data was obtained, a database was constructed without any specific organisation, i.e. the data were processed in the order in which they were obtained. Due to the fact that the literature recommended a treatment of the information, four normalisation measures were applied, which allowed the data to be ready for the next phase of operation. The normalisations used were standard normalisation, uniform normalisation, and granulation (coarse-grained and fine-grained). Based on the literature, this was a procedure that needed to be implemented.

Once the information was standardised, we proceeded to evaluate the precision, sensitivity, accuracy, specificity and F1 score of four of the most widely used classification models in the literature, which were KNN, K means, support vector machines and decision trees. With the results obtained, it could be preliminarily concluded that there is a model that stands out in its level of accuracy as is the KNN, but that there is very little difference between the four models and that it can be at the choice of the expert that one model or another can be used. For the correct application of the models, three measures of similarity were tested, such as the cosine distance, the Jaccard coefficient and the Sorence-Dice coefficient, which were tested and their results were studied with the Friedman and Friedman statistic improved by Holm, finding that the cosine distance was the most appropriate for the scope of this research.

On the other hand, and to further enhance this research, an experimental phase was carried out to test the statistical accuracy of the selected clustering techniques, which allowed us to observe that indeed the KNN model obtains the best results compared to the rest. Furthermore, we can conclude that this model (and very closely the rest) manages to adequately satisfy the tastes and preferences of most of the study subjects, giving preliminary evidence that the research fulfilled the proposed scope.

To continue, the Mamdani fuzzy model is realised for each of the three classes defined in which the eight variables were grouped. In this way, the thesis argues that evaluating each variable once it has been classified on the basis of its nature and by means of a fuzzy model, allows the behaviour of the values to be controlled so that the decisions that are established on the basis of the consideration of this information are within the established ranges and in accordance with the rules of the fuzzy engine.

As explained in the paper, each fuzzy variable becomes a new input variable to the neural network. The backpropagation model therefore appears to be adequate as it stabilises at a performance close to 80%. The experiments were conducted to compare decisions made by a traffic light programmed in the traditional way by traffic experts against a neuro-fuzzy green-light timing control system, achieving with these results establishes that the general objective of the research is fulfilled and responds directly to the assumptions made in relation to the achievement obtained.

In this way, the thesis reports the evaluation of the performance of four clustering techniques, the generation of the membership functions and their alignment to generate input data to a neural network, so that a neuro-fuzzy model is built to solve some traffic problems such as large volumes of cars at intersections generated by a synchronisation of the traffic light phases not adequate to the demands of the moment, which was achieved efficiently and accurately, showing interesting and open results for future research.

6.2 Future Work

The study of vehicle traffic control is a situation that has aroused great interest for many years because it leads to many situations, among which we can highlight the following:

- The growth in demand for service grows based on the number of cars on the roads and this is something that cannot be controlled and is sought to be closely estimated by means of gauging, surveys, etc.
- The way in which users drive is another interesting aspect to evaluate, as despite the regulations and controls in place, people do this activity in an uncertain way and modelling behaviour is a very interesting challenge, but with a high level of complexity.
- Finding the technique that could be the most suitable for all road challenges seems to be a utopia and rather, we are approaching the moment of establishing that adequate control to create intelligent transport systems requires a large-scale project involving experts in different areas of

computer science and other related problem areas.

In light of these facts, it is important to note that the literature has reported a high level of contributions that demonstrate very significant advances, but most of these remain as proposals without real experimentation. This is understandable since the big challenge is not the implementation as such, but to achieve a paradigm shift in the thinking of city administrators and above all of the people living in the city, who at the end of the day, will be the real users of the system. Because of this, the thesis proposes that it is necessary to continue investigating and providing efficient and robust solutions in this important area of study, mainly addressing the following two areas of opportunity:

1. Modelling the behaviour of cars within an intersection so that there is a control to anticipate what might happen on that road, all based on the type of users and the way they drive.
2. Find a methodology that is robust, open and comprehensive enough to include all aspects that are actually active on a road, for example:
    a) Pedestrians
    b) Cars with different speeds, destinations, etc.
    c) Affectations to a road.

In this way, it is proposed to conduct work that is capable of providing solutions in these areas and has the opportunity to take aspects of this research as an important starting point.

6.3 Communications and Contributions

Adan Acosta M., Jose A. Castan R., Salvador Ibarra M., Jesus D. Teran V., Julio Laria M. y Mayra G. Trevino B., **"Evaluation of Data Mining Techniques in Urban Traffic Problem"**, *Journal of Computer and Communication, ISSN: 2327-5227.* Por aparecer en Octubre.

Adan Acosta Morales, **"Modeling of Variables in the Urban Traffic Problem"**, *2nd*
International Graduate Research Colloquium, 2019.

Adan Acosta Morales, **"Propuesta neuro-difusa para mejorar las condiciones del**
Urban Traffic", *1st International Colloquium on Postgraduate Research,* 2018.

References

(Augsten y Bohlen 2013) Augsten, N., & Bohlen, M., Similarity Joins in Relational Database Systems. Morgan & Claypool Publishers, 2013.

(Balbo and Pinson, 2005) Balbo, F., and Pinson, S. 2005. Dynamic modeling of a disturbance in a multi-agent system for traffic regulation. Decision Support Systems, 41:131-146. [doi:10.1016/j.dss.2004.06.001]

(Belikov et al. 2013) D.A. Belikov, S. Maksyutov, V. Sherlock, A. Aoki, N. M. Deutsceher, S. Dohe, D. Griffith, E. Kyro, I. Morino, T. Kakasawa, J. Notholt, M. Rettinger, M. Schneider, R. Sussmann, G. C. Toon, P. O. Wennberg and D. Wunch, Simulation of column-average CO2 and CH4 using the NIES TM with a hybrid sigma-isentropic (a-0) vertical coordinate, Atmospheric, Chemistry and Physics, European Geoscience Union, doi: 10.5194/acp-13-1713-2013.

(Borne et al., 2003) Borne, P., Fayech, B., Hammadi, S., et al., 2003. Decision support system for urban transportation networks. IEEE Transactions on Systems, Man and Cybernetics. Part C: Applications and Reviews. 33(1), 67-77. [doi: 10.1109/TSMCC.2003.809355].

(Bond y Gasser, 1988) Bond A. Y Gasser L., Readings in Dsitributed Artificial Intelligence. Ed. Morgan Kaufmann. ISBN: 978-0934613637. Pp. 649. Agosto, 1988.

(Butler et al., 2012) Butler, T. M., Stock, Z. S., Russo, M. R., Denier van der Gon, H. A. C., and Lawrence, M. G.: Megacity ozone air quality under four alternative future scenarios, Atmos. Chem. Phys., 12, 4413-4428, doi:10.5194/acp-12-4413-2012, 2012.

(Calzada et al., 2017) Calzada-Ledesma, V., Puga-Soberanes, H. J., Rojas-Dominguez, A., Ornelas-Rodriguez, M., Carpio-Valadez, J. M., & Gomez-Santillan, C. G., Comparing Grammatical Evolution's Mapping Processes on Feature Generation for Pattern Recognition Problems. In Nature-Inspired Design of Hybrid Intelligent Systems. Mexico: Springer International, 2017.

(Chen et al., 2006) Chen, B., Cheng, H. H. and Palen, J. 2006. Mobile-C: A mobile agent platform for mobile C/C++ agents. Journal of Software: Practice and Experience 36, 1711-1733. [doi: 10.1002/spe.742]

(Chen y Cheng, 2010) Chen, B. and Cheng, H. 2010. A Review of the Applications of Agent Technology in Traffic and Transportation Systems, IEEE Transactions on Intelligent Transportation Systems, Vol. 11, No. 2. [doi:10.1109/TITS.2010.2048313]

(Cohen and Manion, 2002) Cohen, L., and Manion, L., Metodos de Investigacion Educativa. Madrid, Spain: La Muralla, 2002.

(d'Inverno and Luck, 2001) M. D'Inverno and M. Luck. "Understanding Agent Systems," 2nd Edition, Springer-Verlag, 2004.

(Dautenhahn and , 02) K. Dautenhahn y C. Nehaniv, Imitation in Animal and Artifacts, MIT Press, ISBN: 0262042037, pp. 607, 2002.

(David y Guillermo, 2018) David, W., & Guillermo, G. (2018). Pilot Support System: A Machine Learning Approach, 325-328. https://doi.org/10.1109/ICSC.2018.00067.

(Fish, 2006) B. Fish, "Project to Decrease Air Pollution from Automotive Vehicles in Mexico City by Improving Public Transport and Investing in Clean Technologies", MILAGRO Project (Megacity Initiative: Local and Global Research Observation), Atmospheric, Chemistry and Physics, European Geoscience Union.

(Gaikwad y Borgiri 2015) Gaikwad, S., & Bogiri, N, Levenshtein distance algorithm for efficient and effective XML duplicate detection. IEEE International Conference on Computer, Communication and Control (IC4-2015).

(Garber et al., 88) Garber, Nicholas J. And Hoel, Lester A. Traffic and Highway Enginnering, West Publishing Company, St. Paul, Mn, 1988.

(Garcia-Serrano et al., 2003) Garcia-Serrano, A. M., Teruel, D., Carbone, F., et al., 2003. FIPA-compliant MAS development for road traffic management with a knowledge-based approach: The TRACK-R agents. In Proc. of Challenges Open Agent Syst. Workshop, Melbourne, Australia, 2003.

(Heinemann 2003). Heinemann, K., Introduccion a la Metodologia de la Investigacion Empirica en las Ciencias del Deporte. Editorial Paidotribo, 2003.

(Huang et al., 2012) Huang, S., Sadek, A. and Zhao, Y. 2012. Assessing the Mobility and Environmental Benefits of Reservation-Based Intelligent Intersections Using an Integrated Simulator, IEEE Transactions on Intelligent Transportation Systems.

(Ibarra et al., 2014) S. Ibarra, J. A. Castan y J. Laria, "Optimizaing urban traffic control using a rational agent", En Journal of Zhejiang University Science C, Vol. 15, Issue 12, pp. 1123-1137, doi: 10.1631/jzus.C1400037, 2014

(Jennings y Bussmann, 03) N. R. Jennings and S. Bussmann. "Agent-Based Control Systems. Why Are They Suited to Engineering Complex Systems?," IEEE Control Systems Magazine, vol. 23, no. 3, pp. 61-73, Jun. 2003.

(Jiawei et al., 2006) H. Jiawei, M. Kamber y J. Pei, "Data mining: concepts and techniques", itd: Morgan Kaufmann Vol. 5, pp. 284-304, 2006

(Joh 1997) Joh, D., CBR in Changing Environment. Case Based Reasoning Research and Development. Providence, IR, USA: ICCBR-97, 1997.

(Khadilkar, 2018) Khadilkar, H. (2018). A Scalable Reinforcement Learning Algorithm for Scheduling Railway Lines, 1-11.

(Kozlak et al., 2007) J. Kozlak, Multi-Agent Environment For Modelling And Solving Dynamic Transport Problems, Computing and Informatics, Vol. 28, 2009, 277-298.

(Little et al., 1981) Little, J. D. C., M. D. Kelson and N. H. Gartner, MAXBAND: A Program for Setting Signals on Arteries and Triangular Networks, Transportation Research Record 795, pp: 40-46. 1981.

(Murtagh y Saunders 1993) Murtagh B A, Saunders M.A., MINOS 5.4 user's guide. Technical report, SOL 83-20R, Systems Optimisation Laboratory, Department of Operations Research, Stanford University, Stanford, CA, 1993.

(Mystowski y Khan, 1998) C. Mystowski y S. Khan, Estimating Queue Lengths Using SIGNAL94, SYNCHRO3, TRANSYT-7F, PASSER II-90, and CORSIM, In Proceedings of the 78th Transportation Research Board Annual Meeting, November, 1998.

(Ossowski et al., 2005) Ossowski, S., Hernandez, J. Z., Belmonte, M.V., et al., 2005. Decision support for traffic management based on organisational and communicative multiagent abstractions. Transportation Research Part C: Emerging Technologies, 13(4), 272-298. [doi: science/article/pii/S0968090X0500032X].

(Perez et al., 2010) Perez, J., Seco, F., Milanes, V., et al., 2010. An RFID-Based Intelligent Vehicle Speed Controller Using Active Traffic Signals, Sensors 10(6),5872-5887. [doi:10.3390/s100605872]

(Rodriguez et al., 2001) Rodriguez, M., Alvarez, S., & Bravo, E., Coeficientes de Asociacion. Mexico: Plaza y Valdes, 2001.

(Roozemond, 2001) Roozemond, D. A. 2001. Using intelligent agents for pro-active, realtime urban intersection control. European Journal of Operational Research 131(2), 293301. [doi: science/article/pii/S0377221700001296]

(Saeed et al., 2011) Saeed Y, Khan, S., Ahmed, K., et al., 2011. A Multi-Agent Based Autonomous Traffic Lights Control System Using Fuzzy Control, International Journal of Scientific & Engineering Research Volume 2, Issue 6, June-2011 1, ISSN 2229-5518.

(Sanchez and Alanis 2006) Sanchez E. and Alanis A., Redes neuronales: conceptos fundamentales y aplicaciones al control automatico, Ed. Pearson de Mexico, ISBN: 978848-3222959, pp. 210, 2006.

(Little et al., 1981b) Little, John D. C.; Kelson, Mark D.; Gartner, Nathan H., MAXBAND: A Versatile Program for Setting Signals on Arteries and Triangular Networks, Publisher: Alfred P. Sloan School of Management,

Massachusetts Institute of Technology, 1981.

(Liu et al., 2005) Liu, Z. Q., Ishida, T. and Sheng, H. Y. 2005. Multiagent-based demand bus simulation for Shanghai. In Proc. Massively Multi-Agent Syst. I, vol. 3446, pp. 309-322, 2005.

(Luck et al., 2005) M. Luck, P. McBurne, O. Shehory and S. Willmott. "Agent Technology: Computing as Interaction," A Roadmap for Agent Based Computing, Compiled, written and edited by M. Luck, P. McBurney, O. Shehory, S. Willmott and the AgentLink Community, pp. 11-12, 2005.

(Vieira et al., 2009) Vieira, L. P., Ortiz, L. I., & Segundo, S., Introduccion a la Mineria de Datos. Brazil: E-papers, 2009.

(Wallace et al., 1991) Wallace, Charles, Chang, Edmond, Messer, Carroll, and Courage, Kenneth, Methodology for Optimizing Signal Timing: PASSER II-90 Users Guide, Office of Traffic Operations and Intelligent Vehicle/Highway Systems - Federal Highway Administration, Volume 3, December 1991.

(Wang, 2005) Wang, F. Y. 2005. Agent-based control for networked traffic management systems. IEEE Intelligent Systems 20(5), 92-96. [doi: 10.1109/MIS.2005.80]

(Wang, 2008) Wang, F. Y. 2008. Toward a revolution in transportation operations: AI for complex systems. IEEE Intelligent Systems 23(6), 8-13. [doi: 10.1109/MIS.2008112]

(Webster, 1966) Webster, F.V. and B.M. Cobbe. Traffic Signals, Road Research, Technical Paper No. 56, Her Majesty's Stationery Office, London, England, 1966.

(Weiss, 1999) G. Weiss. "Multi-agent Systems: A Modern Approach to Distributed Artificial Intelligence," Edited by Gerard Weiss. ISBN: 0-262-23203-0, 1999.

(Wooldridge & Ciancarini, 1999) M. Wooldridge and P.Ciancarini. Agent-Oriented Software Engineering: The State of the Art In P. Ciancarini and M. Wooldridge, editors, Agent- Oriented Software Engineering. Springer-Verlag Lecture Notes in AI Volume 1957, January 2001.

(Wooldridge, 2002) M. Wooldridge. "An Introduction to Multi-agent Systems," Published in February 2002 by John Wiley & Sons (Chichester, England), ISBN: 0 47149691X, 2002.

(Zade and Dandekar, 2011) Zade, A. R. and Dandekar, D. R. 2011. FPGA Implementation of Intelligent Traffic Signal Controller Based On Neuro-Fuzzy System. International Conference on Advanced Computing, Communication and Networks.

(Zhang et al., 2004) Zhang, H. S., Zhang, Y., Li, Z. H., et al., 2004. Spatial-temporal traffic data analysis based on global data management using MAS.

IEEE Transactions on Intelligent Transportation Systems, (4)267-275. [doi: 10.1109/TITS.2004.837816].

Printed by Books on Demand GmbH, Norderstedt / Germany